Pocket Guide to the

Bumblebees

of Great Britain and Ireland

Richard Lewington

BLOOMSBURY WILDLIFE

LONDON · OXFORD · NEW YORK · NEW DELHI · SYDNEY

Dedicated to my daughter Alexandra and
to my granddaughter Florence

HELM
Bloomsbury Publishing Plc
50 Bedford Square, London, WC1B 3DP, UK
29 Earlsfort Terrace, Dublin 2, Ireland

BLOOMSBURY, HELM and the Helm logo are trademarks of Bloomsbury Publishing Plc

First published in the United Kingdom 2023

A catalogue record for this book is available from the British Library

Library of Congress Cataloguing-in-Publication data has been applied for

ISBN: PB: 978-1-4729-9359-5; ePUB: 978-1-4729-9361-8; ePDF: 978-1-4729-9360-1

2 4 6 8 10 9 7 5 3 1

Typeset and designed by D & N Publishing, Baydon, Wiltshire
Printed and bound in Italy by L.E.G.O. SpA

Contents

Acknowledgements

I would like to thank the following people for their help, enthusiasm and constructive comments during the preparation of this book: Dr Richard Comont, Steven Falk, Professor Dave Goulson, Gill Perkins, Royal Mail Group Limited and Amoret Spooner OUM. Thanks also to David Campbell and Katy Roper of Bloomsbury, and Susi Bailey and David Price-Goodfellow for their meticulous attention to detail in copy-editing and design, respectively.

Queen Common Carder
Bee on a Pea flower

Foreword

Bumblebees have always represented summer to me through their wonderful droning buzz as they bumble from flower to flower: the sound of summer, played out in our cities, towns, gardens and countryside. Bumblebees are not just beautiful but are also fascinating to watch.

Bees, of course, play a vital role in our lives. They make it possible for us to grow fruit and vegetables, and to enjoy thriving, colourful parks and gardens. The wider natural landscape would look very different without them, and our diets, health and lives would also be less rich – our food scarce and more expensive, with much less variety.

Becoming the chief executive officer of the Bumblebee Conservation Trust (BBCT) has been a journey of discovery and gratefulness. My father instilled in me a love of nature, and I will be forever thankful for his dedication in showing me a world I may not have discovered by myself. My passion to find a career 'outside' led me to study ecology but also led me into farming. Here, the connection between how our food is produced and our dependency on pollinators to provide that food was cemented, and bees – bumblebees – became an all-consuming enthusiasm.

I first met Richard Lewington six years ago when he kindly provided his stunning artwork for an event the BBCT ran in London. He is without doubt one of Europe's leading wildlife artists, and he includes the brilliant *Field Guide to the Bees of Great Britain and Ireland* (2015), written by Steven Falk, among his many previous books.

In this new pocket guide, Richard exemplifies the beauty of all the bumblebees that occur naturally in Britain, Ireland and the Channel Islands. His superb drawings of bumblebees visiting flowers illustrate what everyone can see in their gardens, on walks in the countryside or in towns. We can all benefit from understanding how our rarest bees can be identified; the Shrill Carder Bee, Great Yellow Bumblebee, Moss Carder Bee and Brown-banded Carder Bee need our help more than any other species, and the accompanying detailed text will help us identify and treasure them, and learn how we can help to reverse their recent declines.

I am confident this book will help all understand and value these precious creatures and give us an enduring affection for them.

Gill Perkins
CEO, Bumblebee Conservation Trust

Conservation

The 20th century saw the industrialisation of farming, with the introduction of mechanisation, synthetic fertilisers and an ever-growing plethora of pesticides. During and following the Second World War, government policies encouraged this process. Small farms were swallowed by larger ones, and small fields were merged into bigger ones. Approximately half of our hedgerows were destroyed between 1950 and 2000. Ninety-seven per cent of our flower-rich hay meadows and chalk downland was lost, along with 80 per cent of our lowland heaths. Flower-rich habitats were replaced with monocultures of a small number of arable crops, or fields of 'improved' rye grass for silage and pasture. All of these changes made our countryside far less hospitable to bumblebees, with fewer quiet places for them to nest and far fewer flowers.

Much debate and controversy has focused on the importance of pesticides in wild bee declines, particularly neonicotinoid insecticides. Ever since DDT was introduced in the 1940s, a pattern has emerged in which new pesticides are initially judged to be safe and used by farmers for several decades, until evidence emerges that they are harming the environment, when they are banned and replaced with a new product. Neonicotinoids were the most recent type of pesticide to follow this pattern. They were introduced in 1994, but most were banned in Europe in 2018 when evidence emerged that these potent neurotoxins were contaminating soils and the nectar and pollen of wildflowers, leading to poisoning of bees. Many of their effects were 'sub-lethal': exposed bees have reduced memory and navigation skills, lower fertility and low resistance to disease. It remains to be seen whether new generations of insecticides will prove to be any safer. Overall, the frequency of pesticide applications to crops continues to increase, with UK crops receiving more than 16 applications per year on average, up 70 per cent compared to 1990. As a result, wild pollinators living in or near farmland are chronically exposed to complex mixtures of pesticides. It seems very likely that this continues to contribute to bumblebee declines.

If we are to halt and reverse bumblebee declines, then a number of different actions are necessary. Collectively, gardens and other urban green spaces such as road verges, roundabouts, parks and so on could provide a national network of flower-rich habitats (see opposite and overleaf). We also need to find ways to restore more flowers to the countryside, since farmland covers about 70 per cent of the British Isles. The loss of flower-rich grasslands such as hay meadows was particularly important in driving declines of a suite of bumblebee species that were strongly associated with this habitat, including the Shrill Carder Bee, Brown-banded Carder Bee, Ruderal Bumblebee and Great Yellow Bumblebee. Cullum's Bumblebee, a species found on chalk downland, went extinct in the 1940s, while the Short-haired Bumblebee followed it to extinction in the 1980s. These species all tend to emerge from hibernation late in spring, to coincide with the flowering of clovers and other legumes in open grassland. They also tend to have longer tongues and prefer deep flowers such as Red Clover, Tufted Vetch, Kidney Vetch, Common Knapweed and Yellow Rattle, all plants associated with grasslands. If these species are to recover, we must restore or re-create more of these beautiful meadows.

Re-creation of flower-rich meadows is not simple, however, and sometimes takes many years. It is unlikely to reach the high level of species richness found in an ancient meadow, but nonetheless can be very successful in supporting a high diversity of insects and other wildlife. It is easiest on soils with low fertility, since otherwise fast-growing weedy plants such as nettles and docks tend to take over, outcompeting wildflowers. The legumes that are loved by many of our rarer bumblebees tend to thrive in poorer soils where their ability to fix nitrogen from the atmosphere with the help of symbiotic bacteria that live in nodules in their roots gives them an advantage. On soils that have recently been treated with artificial fertilisers, it is often wise to grow and harvest a crop such as oilseed rape, without applying any fertiliser, to reduce soil fertility before sowing any wildflower seeds. A more extreme and expensive option is to scrape off some of the topsoil. It is best to find a source of wildflower seeds from as close as possible to the site, so that they are of local provenance, adapted to the local soil and climate. Another option is to spread green hay harvested from a nearby flower-rich meadow, if one exists and the owners are willing. More details on restoring or creating wildflower meadows can be obtained from the Bumblebee Conservation Trust (BBCT).

An alternative approach to sowing wildflower mixes is to allow natural regeneration to occur, via seeds in the soil or blown in on the wind. The exciting new 'rewilding' projects popping up around the UK are essentially using this approach, stepping back and allowing nature to take its course. An integral part of rewilding is including large herbivores to provide grazing and disturbance, mimicking the situation that prevailed thousands of years ago before humans came to dominate the landscape. An example is the Knepp Wildland project in West Sussex, which has been running since 2001 and now provides many more wildflowers than were there when the landscape was intensively farmed. The area also supports many more insects, including some rarities, but it has not yet been colonised by any rare bumblebee species.

A key issue in conserving bumblebees is providing sufficient high-quality connected habitat to support viable populations in the long term. Bumblebees need substantial areas of good habitat since their populations are determined by the number of nests (each containing just one breeding female). Small fragments of hay meadow or downland, or isolated gardens full of flowers, are therefore of little use to them unless they are well connected to many other small fragments. At present there simply isn't enough connectivity. The charity Buglife has long recognised the need for increased connectivity, and its B-Lines project is attempting to create corridors of flower-rich habitat criss-crossing the UK.

Since farming is the dominant form of land use in Britain and Ireland, the most promising approach is to help and encourage farmers to incorporate more flower-rich habitat and reduce pesticide use, making farmland more welcoming to a diversity of wildlife – including bumblebees. This would help to link up nature reserves and other areas of high nature value. Financial support for farmers to sow and manage wildflower strips along field margins has been in place for many years, but take-up has been small. Funding is also available for re-creating meadows and planting more hedges. At the time of writing, the system of support for agri-environment schemes is being revised by the UK government, and new Environmental Land Management schemes are being introduced. These will hopefully offer more support for farmers to move towards more sustainable, wildlife-friendly practices. However, there seem to be no government initiatives aimed at reducing pesticide use.

The BBCT was formed in 2006 to bring about an awareness of bumblebees and their importance, not only as pollinators, but also as an integral part of the sounds and sights of the British and Irish countryside. A science-led organisation, its work ranges from education and encouraging the commoner species to thrive in garden and urban areas, to the management of more sensitive habitats for some of our rarer and endangered bumblebees. Further information on how you can become involved and learn more about bumblebees can be obtained from: www.bumblebeeconservation.org.

Dave Goulson
Professor of Biological Sciences, University of Sussex,
and founder of the Bumblebee Conservation Trust

Introduction

Bumblebees are perhaps the most familiar members of the huge and diverse order of insects known as Hymenoptera, which includes sawflies, ants, bees and wasps. In Britain and Ireland there are more than 6,700 species of Hymenoptera, of which about 270 are bees. Of these, only 27 species of bumblebee have been recorded. Three of them – the Apple Bumblebee, Cullum's Bumblebee and Short-haired Bumblebee – have been declared extinct in the British Isles, although the Short-haired Bumblebee is currently undergoing a reintroduction process following its disappearance more than 30 years ago. Worldwide, there are about 250 bumblebee species. They are found mainly in the northern hemisphere, suggesting that these well-insulated insects are rather more tolerant of cooler temperatures than are other less hairy bees.

Of the 24 remaining species of resident bumblebees, six are known as cuckoo bumblebees, with behaviour mimicking that of the Cuckoo bird (*Cuculus canorus*), which parasitises the nests of other species. Cuckoo bumblebees take over the nests of similar-looking true bumblebee species and can sometimes be difficult to distinguish from their hosts. Further details on separating cuckoo bumblebees from true bumblebees can be found on page 14.

In the British Isles, one has only to look at the way in which the media has hijacked the image of the bumblebee to judge how universally well loved these endearing insects are. Their rotund, fluffy bodies bumbling from flower to flower on warm, sunny days epitomise midsummer. However, they are more than just cuddly-looking insects, telling us that spring has arrived. Bumblebees are essential in gardens and the wider countryside as pollinators, visiting the flowers of a huge variety of plants, from fruit trees, soft fruits and vegetables to ornamental plants, wildflowers, trees and shrubs. Indeed, they are part of an army of insects regarded as being responsible for one-third of all the insect-pollinated food humans consume. Listed on pages 93 and 94 are native and non-native plants regularly visited by foraging bumblebees.

Many other familiar insects also act as pollinators, including flies, wasps, beetles, butterflies and moths, but bumblebees are particularly efficient at the task because all their time is spent visiting flowers and their hairy bodies gather and inadvertently transfer large amounts of pollen. Although the majority of bumblebees directly transfer pollen from one flower to another, some species – particularly those with short tongues – may bite into the back of long flower tubes and steal nectar without collecting pollen in the process. This is known as 'nectar robbing'.

The closely related, slimmer and less hairy Western Honey Bee (*Apis mellifera*) is famous for its role as a pollinator, but per bee it is far less efficient than are bumblebees. That said, what it lacks in efficiency it makes up for in sheer numbers, with hives often containing several tens of thousands of workers, compared to an average of a few hundred workers in a bumblebee nest. This can lead to introduced honeybees outnumbering and outcompeting native bees and other pollinators when resources are limited. This fact should be remembered when positioning hives, and sites important for bumblebees and other pollinators, which are far less able to compete with honeybees, should be avoided.

Although Britain and Ireland's bumblebees have been well studied since the publication of Frederick Sladen's expansive *The Humble-bee: Its Life-history and How to Domesticate It* in 1912, there is still much to learn and new information is constantly being acquired. Recording behaviour and distribution is essential if we are to help and manage habitats for bumblebees, but before that it is necessary to correctly identify the species we encounter. As mentioned above, we presently have just 24 resident species on the British and Irish lists – or 27 if we include the two extinct species and one reintroduction. Identification of these species in the early months of the year, when freshly emerged queens are on the wing, is relatively straightforward. However, later in the year, when workers and the often variable males have appeared, things may become rather more complicated. Many books use photography to help identify bumblebees, but the vagaries of this medium, with its differing angles, lighting and equipment, combined with the varying condition of specimens, can lead to confusion. Often the smallest of details are crucial for correct identification. The aim of this book is to help the reader identify bumblebees using artwork, showing any finer points as clearly and in as much detail as possible, and with each species viewed from exactly the same angles to allow direct comparison.

How to use this guide

The first step in identifying bumblebees is to become familiar with their basic anatomy and structure. Then, learning how to separate cuckoo bumblebees from true bumblebees and how to differentiate between the sexes will take you a step closer to positive identification. The 'At-a-glance guide' on pages 22–25 shows bumblebees grouped according to tail colour and is intended as a quick erence guide. Identification may then be confirmed by following the cross ference to the appropriate species account in the main portion of the book.

Also included in the species accounts are simple diagrams of the face lengths of each bumblebee species. These are categorized as short, medium or long, and can often be useful guides to separate other similar species, for example the Garden Bumblebee *Bombus hortorum*, has the longest face of all our bumblebees, while the Heath Bumblebee, *Bombus jonellus*, has a very short face.

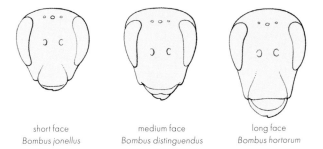

short face	medium face	long face
Bombus jonellus	*Bombus distinguendus*	*Bombus hortorum*

Following some species accounts there are spreads showing similar confusion species, grouped together for direct comparison. If the identification of a specimen still cannot be decided on using the colour illustrations, additional diagrams showing finer details such as male genitalia and antennal segments have been included. In such cases, microscopic examination of dead specimens might be necessary for scientific purposes.

The maps show the general distribution of the species in the British Isles. The phenology charts indicate when queens, workers and males are on the wing, but the extreme dates are likely to vary with the prevailing weather conditions.

Structure and anatomy

The basic structure of a bumblebee follows that of most other insects, being divided into head, thorax and abdomen. The features that help identify and separate bumblebees from similar insects, some of which mimic them for their own protection, are described below. It is helpful to become familiar with the various anatomical terms (see also the Glossary on page 91). Features of particular note are the antennae, which in

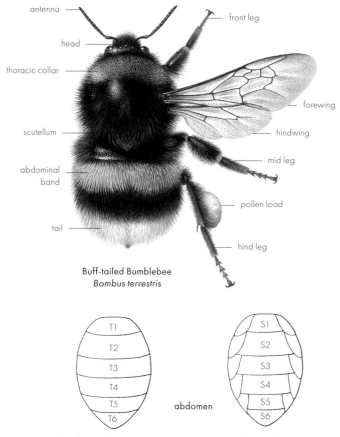

antenna

head

thoracic collar

scutellum

abdominal band

tail

front leg

forewing

hindwing

mid leg

pollen load

hind leg

Buff-tailed Bumblebee
Bombus terrestris

| T1 |
| T2 |
| T3 |
| T4 |
| T5 |
| T6 |

abdomen

| S1 |
| S2 |
| S3 |
| S4 |
| S5 |
| S6 |

dorsal view, showing tergites

ventral view, showing sternites

Hymenoptera are well developed and 'elbowed' at the scape. Bumblebees also have two pairs of wings, the forewings and hindwings, which are linked by a row of tiny hooks. These features distinguish them from some of the often very similar hoverflies (Diptera), which also have shorter, simpler antennae and only one pair of functional wings (see pages 86–90).

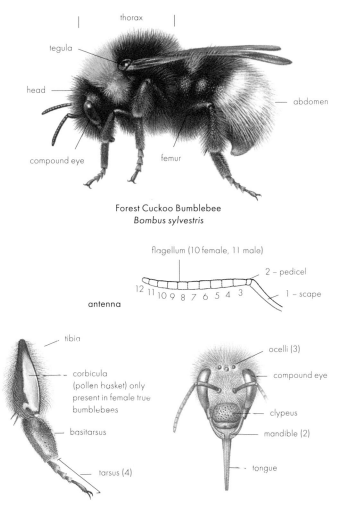

Forest Cuckoo Bumblebee
Bombus sylvestris

flagellum (10 female, 11 male)

2 – pedicel

1 – scape

antenna

tibia

corbicula
(pollen basket) only
present in female true
bumblebees

basitarsus

tarsus (4)

hind leg, lateral view

ocelli (3)

compound eye

clypeus

mandible (2)

tongue

head, front view

True bumblebee or cuckoo bumblebee?

As mentioned earlier, our region includes six cuckoo bumblebee species, which parasitise true bumblebees. Cuckoo bumblebees – especially the males – are often less active than true bumblebees, and their buzz has a deeper, quieter hum. They have broader, square-shaped heads and are often more powerfully built than their hosts. Their shiny exoskeleton shows through the sparsely hairy body, and females in particular have darker wings. Below are other ways of separating members of the two groups.

pollen basket
on hind tibia shiny;
flat or slightly concave

mandibles
square

Garden Bumblebee
Bombus hortorum
(pages 34–35)

True bumblebees are densely hairy, making them less shiny than cuckoo bumblebees, and their wings are never very dark. The male hind legs are similar to, but more slender than those of the female, and the tibia are less shiny.

hind tibia
hairy, slightly
convex

mandibles
triangular

Southern Cuckoo Bumblebee
Bombus vestalis

Cuckoo bumblebees have sparser body hair, giving them a shiny appearance. They are more robust in build, with a broad, box-shaped head. Males have a blunt, tapered abdomen. No workers are produced, just females and males. All cuckoo bumblebees have short to medium-length tongues.

Bumblebee – female or male?

The first thing to determine when identifying bumblebees is whether an individual is a female or a male. Males appear slightly later in the year, so large bees found in early spring will be queens. The male is usually more slender than the female and his antennae are noticeably longer. If pollen loads are being carried, then the bee is a queen or worker. Only the females are capable of stinging, although they rarely do so. Unlike honeybees, the bumblebee's sting is smooth, not barbed, so the bees do not die after they have stung as the sting can be extracted. Below are other ways of separating the sexes.

underside tip of
abdomen pointed

antenna,
12 segments

Female White-tailed
Bumblebee
Bombus lucorum

Female bumblebees usually have rotund, furry bodies with six abdominal segments and a pointed tail containing the sting. The hind tibia of true bumblebees is concave and shiny, and fringed with long hairs for carrying pollen. Queens are generally larger than the sterile workers.

underside tip of
abdomen rounded

antenna,
13 segments

Male White-tailed
Bumblebee

Male bumblebees are usually smaller and fluffier than queens, and they often have more pale yellow hairs, especially on the face and head. They have seven abdominal segments and a blunt, rounded tip to the tail. Usually less energetic than workers, they do not gather pollen, although pollen often collects on their fur as they feed. The tibia is hairy.

15

The life cycle of a true bumblebee

The main task of a queen bumblebee on her emergence in spring after hibernation is to find a mate. Only the new queen survives through the winter and into the following year – the males and workers all die after a fairly short life (although unlike male honeybees, male bumblebees don't die as a result of mating).

Towards the end of summer, a queen bumblebee will seek a sheltered spot away from the coming winter's sun, perhaps in a cool bank or under bark, and here she will excavate a chamber in which to hibernate. She emerges the following spring as the temperature rises. This may be as early as February or March, but queens of some species wait until May or June before appearing. When her body temperature has risen, the queen embarks on a foraging flight in search of a feast of pollen and nectar, which is essential to build up her strength and to enable her ovaries to mature before nest-making and egg-laying begin.

After several days of feeding activity and when fully replenished, the queen begins her slow, purposeful flight in search of a suitable nesting site, inspecting nooks and crannies and often disappearing below ground and then reappearing after several minutes. She is a fussy homemaker and may spend many days searching before finding the right place. Frequently, a queen bumblebee will choose an old vole or mouse nest, on or just below the surface of the soil. She often uses the remains of grass and moss from the old rodent nest to form a chamber with an entrance tunnel, which will expand as the colony grows. Queens of the five species of carder bee often choose to nest on the surface among longer grasses, and rake or 'card' grass and moss over their nests to protect and insulate them.

After selecting a suitable nesting site, the queen then sets off to forage, filling the baskets on her hind legs with heavy pollen loads and taking these back to the nest. Using wax exuded from between the segments on the back of her abdomen, she constructs a cell into which she moulds the pollen clump, moistened with nectar. Depending on the species, the queen lays around 8–16 eggs in the clump, after which she adds more pollen and builds a canopy from a thin mix of wax and pollen. She also uses wax to build a marble-sized nectar pot, which is positioned within easy reach of the pollen clump, allowing her to sustain her energy as she lies on top of the chamber containing the eggs.

After 4–5 days, the warmth of the queen's body, together with the vibrating of her flight muscles to raise the temperature, encourages the development of the eggs. These hatch into young larvae and each one creates a hollow in the bed of the pollen. Growth is rapid, as the larvae are surrounded by an abundance of food, which the queen supplements with a mix of nectar and pollen. Bumblebees that use this method to feed their young are called 'pocket makers', but some more advanced species, known as 'pollen storers', build pots away from the brood or use empty cocoons to hold their collected pollen, and the queen and workers feed the young by regurgitation through the wall of the canopy. After 10–20 days, following several moults, the larvae pupate, each one forming an oval silken cocoon in an upright posture, leaving the old wax chambers for the queen to recycle for successive broods.

About a month after egg-laying, the first new bumblebees, which are all female workers, emerge by biting their way out of their cocoon. After cleaning themselves they take their first drink from the nearby nectar pot. In a day or so their bodies harden and they attain their familiar markings; they are now ready for the outside world and the life of the social colony begins. The new workers set off on their task of collecting pollen and nectar, taking over responsibility for building new cells from the queen as she provisions them with more and more eggs. The queen's outdoor activities are now at an end and she will spend the rest of her life inside the nest, laying eggs. These hatch into more workers and, eventually, males and new queens. The latter leave the nest and mate with a male soon after. They then build up their fat stores in preparation for hibernation. As the weather cools, the old queen dies, as do the workers and males, leaving the new queens to begin the cycle again the following year.

The life cycle of a cuckoo bumblebee

The six British and Irish species of cuckoo bumblebee are all social parasites that favour certain true bumblebee species as their hosts. Some true bumblebees (particularly those that emerge late in the year) will sometimes take over the newly established nests of other queens via a process known as usurpation. In contrast, cuckoo bumblebees (which are best described as specialised kleptoparasites) have females that produce no workers of their own, are incapable of producing wax and have no pollen-collecting equipment on their hind legs. The female cuckoo bumblebee has a similar start to her life as the true bumblebee queen, although her spring begins several weeks later, after her host has already set up home.

Although most cuckoo females resemble their hosts, they are generally larger and more powerful, with a harder exoskeleton and a more potent sting, enabling them to overpower the host. This usually allows the cuckoo female to be successful in conquering the host queen, but if the colony is well established, she may herself be repelled or even killed. Once established in the nest and having dominated the workers, the cuckoo female sets about laying her eggs, using existing cocoons and new cells made from old remnants of wax; she will also eat any of the original host's eggs that she encounters. Her larvae are reared by the host workers in the same way as true bumblebees, although no workers of her own are produced, just males and females. After mating, the males will die, leaving the females to hibernate and continue the cycle into the following year.

Winter-active bumblebees

Since the late 1990s there have been increasingly regular observations of bumblebees foraging throughout the winter months, probably as a result of warmer autumn weather, followed by milder winter temperatures. Most of these observations are of queen Buff-tailed Bumblebees and the occasional Early Bumblebee, but workers and males have also been observed, suggesting the establishment of new colonies created by queens who would normally be in hibernation. Most observations have been in southern England, less often in the north, and most are from urban areas, gardens, parks and quite often from garden centres, where a greater diversity of nectar-bearing plants are available. Most observations are made on warm, sunny days, but the densely hairy bodies of bumblebees help protect them from the cold and occasionally queens may be seen even when there is snow on the ground. Throughout the winter months there are far fewer nectar sources available to bumblebees, and they rely heavily on non-native plants, such as Mahonia species, Winter-flowering Honeysuckle, Winter Jasmine, heathers, crocuses and snowdrops.

The Bumblebee Conservation Trust and BWARS (see page 95) are monitoring records of winter-active bumblebees, and all records, along with photos if possible, should be sent to BWARS.

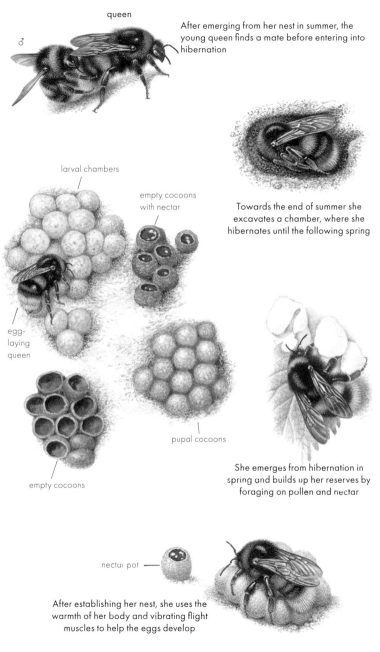

queen

♂

After emerging from her nest in summer, the young queen finds a mate before entering into hibernation

Towards the end of summer she excavates a chamber, where she hibernates until the following spring

larval chambers

empty cocoons with nectar

egg-laying queen

pupal cocoons

empty cocoons

She emerges from hibernation in spring and builds up her reserves by foraging on pollen and nectar

nectar pot

After establishing her nest, she uses the warmth of her body and vibrating flight muscles to help the eggs develop

The 'big seven' garden bumblebees

Gardens are excellent places in which to start learning about bumblebees. Many gardens have a diverse range of flowering trees, shrubs and plants, providing nectar and pollen throughout the year, and as fewer bumblebee species regularly appear in gardens, the identification of those that do is easier. Until the arrival of the Tree Bumblebee (likely from mainland Europe) in 2001 and its subsequent spread, six species were frequently seen in our gardens. Today, that number has risen to the 'big seven'.

White-tailed Bumblebee
Bombus lucorum (pages 26–27)
Females have two bright yellow bands and a pure white tail. Males are quite unlike females, being variable but always with yellow hairs on the head and face.

Buff-tailed Bumblebee
Bombus terrestris (pages 30–31)
Females often have deeper yellow bands and the collar band may be quite dark and narrow. Males and workers always have black hairs on the head and face, and they usually have a buff edge between the white tail and black abdomen.

Garden Bumblebee
Bombus hortorum (pages 34–35)
The double yellow band at the rear of the thorax and base of the abdomen is distinctive. The tongue is longer than in any other species, enabling it to probe long-tube flowers such as Foxglove.

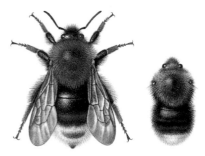

Tree Bumblebee
Bombus hypnorum (pages 40–41)
Males and females have distinctive ginger, black and white markings; the thorax varies from dark russet to light ginger, usually with a darker centre. Some specimens can be almost black but they retain the white tail.

Early Bumblebee
Bombus pratorum (pages 46–47)
Our smallest bumblebee and one of the first to appear in gardens in spring. Workers can be tiny and often lack the central yellow band on the abdomen. The red tail may become faded with age.

Red-tailed Bumblebee
Bombus lapidarius (pages 54–55)
The glossy, jet-black queens have a red tail and are unlike any other common bumblebee. Males are quite different, with extensive yellow markings on the head and thorax, and orange hairs on the hind tibia.

Common Carder Bee
Bombus pascuorum (pages 62–63)
A variable species, typically ginger-buff but often quite pale in northern Britain. The abdomen always has black hairs mixed in, sometimes creating a striped appearance.

At-a-glance guide

This is a quick guide to help identify bumblebees based on tail colour. Consideration should also be given to the age and condition of specimens, as old and sun-bleached individuals may look balder and more faded.

Bumblebees with white or buff tails

White-tailed Bumblebee
Bombus lucorum (pages 26–27)

Buff-tailed Bumblebee
Bombus terrestris (pages 30–31)

Broken-belted Bumblebee
Bombus soroeensis (pages 32–33)

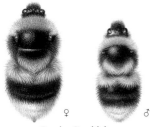

Garden Bumblebee
Bombus hortorum (pages 34–35)

Ruderal Bumblebee
Bombus ruderatus (pages 38–39)

Heath Bumblebee
Bombus jonellus (pages 42–43)

Tree Bumblebee
Bombus hypnorum (pages 40–41)

Short-haired Bumblebee
Bombus subterraneus (pages 52–53)

Cuckoo bumblebees

Forest Cuckoo Bumblebee
Bombus sylvestris (pages 78–79)

Barbut's Cuckoo Bumblebee
Bombus barbutellus (pages 70–71)

Gypsy Cuckoo Bumblebee
Bombus bohemicus
(pages 72–73)

Southern Cuckoo Bumblebee
Bombus vestalis (pages 80–81)

Bumblebees with red tails ×1.5 actual size

Bilberry Bumblebee
Bombus monticola
(pages 44–45)

Early Bumblebee
Bombus pratorum
(pages 46–47)

Red-tailed Bumblebee
Bombus lapidarius
(pages 54–55)

Red-shanked Carder Bee
Bombus ruderarius
(pages 66–67)

Cuckoo Bumblebees

Red-tailed Cuckoo Bumblebee
Bombus rupestris
(pages 76–77)

Bumblebees with ginger or yellow tails

Brown-banded
Carder Bee
Bombus humilis
(pages 58–59)

Moss Carder Bee
Bombus muscorum
(pages 60–61)

Common Carder Bee
Bombus pascuorum
(pages 62–63)

Shrill Carder Bee
Bombus sylvarum
(pages 68–69)

Cuckoo bumblebees

Great Yellow
Bumblebee
Bombus distinguendus
(pages 50–51)

Short-haired Bumblebee
Bombus subterraneus
(pages 52–53)

Field Cuckoo Bumblebee
Bombus campestris
(pages 74–75)

All-black bumblebees

several species of
both true and cuckoo
bumblebees have
non-typical melanic forms

Ruderal Bumblebee
Bombus ruderatus
(pages 38–39)

Field Cuckoo Bumblebee
Bombus sylvestris
(pages 74–75)

White-tailed Bumblebee aggregate

Bombus lucorum, B. magnus, B. cryptarum

Habitat and distribution

Bombus lucorum is commonly found in a wide range of flowery habitats, including urban and brownfield areas. *Bombus magnus* and *B. cryptarum* are more likely to be encountered in upland heaths and moors. The White-tailed Bumblebee aggregate is widespread throughout Britain and Ireland.

The White-tailed Bumblebee aggregate is a combination of three almost identical species: the true White-tailed Bumblebee (*Bombus lucorum*), Northern White-tailed Bumblebee (*B. magnus*) and Cryptic Bumblebee (*B. cryptarum*). They can be reliably separated only using DNA testing, although the extent of the yellow band down the sides of the thorax can help in identification. Because of this uncertainty, sightings should be recorded as '*Bombus lucorum* agg'.

The widespread White-tailed Bumblebee is a mainly lowland species, found in many habitats and frequently occurring in parks and gardens. The other two species are mainly associated with northern moorlands and heaths, although *B. magnus* also occurs on lowland heaths in other parts of Britain and Ireland, notably the New Forest.

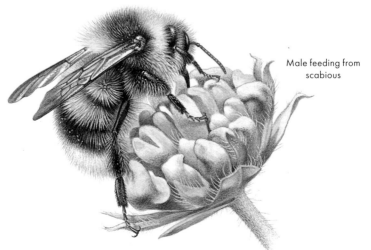

Male feeding from scabious

Nest

Nests are sited under cover, often in old rodent nests, and contain around 200 workers. Two generations may be produced in the south.

Flowers visited

All three species have a short tongue and forage on a wide range of plants. These include willows, dead-nettles and Flowering Currant, and later in the year, knapweeds, Bramble and thistles.

Cuckoo parasites

Gypsy Cuckoo Bumblebee (pages 72–73).

Similar species

See pages 28–29.

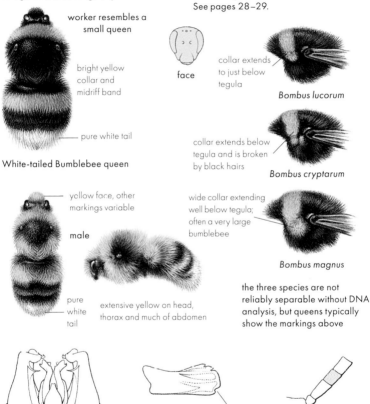

worker resembles a small queen

bright yellow collar and midriff band

pure white tail

White-tailed Bumblebee queen

face

collar extends to just below tegula

Bombus lucorum

collar extends below tegula and is broken by black hairs

Bombus cryptarum

yellow face, other markings variable

male

wide collar extending well below tegula; often a very large bumblebee

Bombus magnus

pure white tail

extensive yellow on head, thorax and much of abdomen

the three species are not reliably separable without DNA analysis, but queens typically show the markings above

male genitalia similar to Buff-tailed Bumblebee

mandible with notch (cf. Broken-belted Bumblebee *B. soroeensis*, page 33)

antennal segment 4 longer than wide (cf. Broken-belted Bumblebee *B. soroeensis*, page 33)

	JAN	FEB	MAR	APR	MAY	JUN	JUL	AUG	SEP	OCT	NOV	DEC
Q												
W												
M												

White-tailed Bumblebee: similar species

Below are illustrated differences between the White-tailed Bumblebee aggregate and other similar species with a yellow collar, single yellow midriff band and white or buff tail.

bright yellow bands, short face

White-tailed Bumblebee
Bombus lucorum

Females

worker

buff band

queen

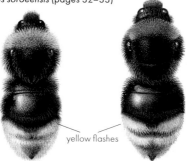

Buff-tailed Bumblebee
Bombus terrestris (pages 30–31)
usually larger, darker, often with narrow collar:
queen has buff tail, worker has white tail

longer face

crescent

buff band

Broken-belted Bumblebee
Bombus soroeensis (pages 32–33)

smaller, yellow midriff usually weakened in centre and extends onto sides of first segment, forming a crescent shape

social parasites that resemble their hosts

sparser body pile, giving a shiny appearance

both lack central yellow midriff bands

outer face of tibia densely haired

yellow flashes

Gypsy Cuckoo Bumblebee
Bombus bohemicus
(pages 72–73)

Southern Cuckoo Bumblebee
Bombus vestalis
(pages 80–81)

28

Males

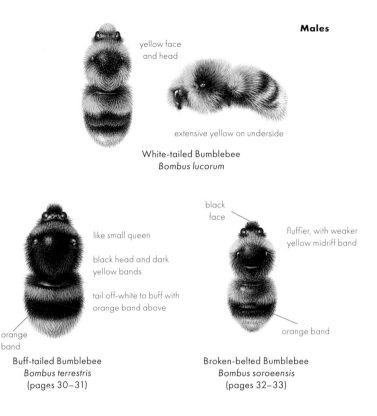

yellow face
and head

extensive yellow on underside

White-tailed Bumblebee
Bombus lucorum

like small queen

black head and dark
yellow bands

tail off-white to buff with
orange band above

orange
band

black
face

fluffier, with weaker
yellow midriff band

orange band

Buff-tailed Bumblebee
Bombus terrestris
(pages 30–31)

Broken-belted Bumblebee
Bombus soroeensis
(pages 32–33)

all three lack a clearly
defined yellow midriff band
and have a shiny abdomen

outer face of tibia densely
haired

Gypsy Cuckoo
Bumblebee
Bombus bohemicus
(pages 72–73)

Southern Cuckoo
Bumblebee
Bombus vestalis
(pages 80–81)

Forest Cuckoo
Bumblebee
Bombus sylvestris
(pages 78–79)

Buff-tailed Bumblebee

Bombus terrestris

Habitat and distribution

Ubiquitous, occurring in most habitats except those in higher mountainous regions. One of the most common and widespread species, found throughout the British Isles. Its range has increased northwards and it has now reached Shetland, although it is still more local in the far north of Scotland.

Huge bumblebees seen in gardens in early spring are almost invariably queen Buff-tailed Bumblebees, one of the 'big seven' common garden species (see page 20). Workers may also be seen on mild winter days in the south, where winter nesting has been confirmed. At this time, they rely heavily for sustenance on winter-flowering shrubs such as *Mahonia*, *Begonia* and *Japonica*. The workers have a white tail and can be almost impossible to separate from White-tailed Bumblebee workers, but a buff edge between the white tail and black abdomen may help separate them.

Controversially, non-native colonies of Buff-tailed Bumblebees have been available commercially for crop pollination since the 1980s, giving rise to concern that these bees could be detrimental to native populations through genetic mixing and disease.

Male feeding on sedum in late summer, showing narrow buff transitional band near tail

Nest
Nests are sited underground and are very large, sometimes containing around 500 workers. They are often made in an old rodent nest and accessed by a sloping tunnel. Two generations or continuous brooding may occur in the south.

Flowers visited
The species has a short tongue and feeds on a wide range of cultivated and wild flowers. In spring these include willows, Blackthorn, gorses, daffodils and crocuses; later in the year, it visits scabiouses, knapweeds, Michaelmas Daisy and sedums. It also engages in robbing nectar from long-tubed flowers.

Cuckoo parasites
Southern Cuckoo Bumblebee (pages 80–81).

Similar species
See pages 28–29.

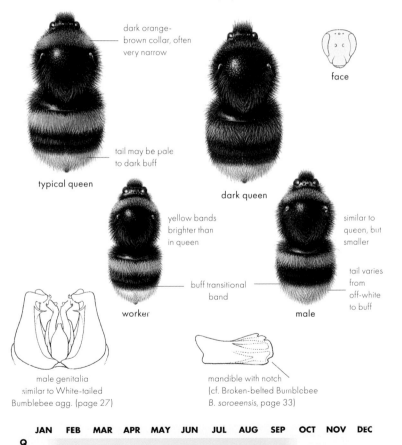

dark orange-brown collar, often very narrow

tail may be pale to dark buff

typical queen

dark queen

face

yellow bands brighter than in queen

buff transitional band

worker

similar to queen, but smaller

tail varies from off-white to buff

male

male genitalia similar to White-tailed Bumblebee agg. (page 27)

mandible with notch (cf. Broken-belted Bumblebee *B. soroeensis*, page 33)

	JAN	FEB	MAR	APR	MAY	JUN	JUL	AUG	SEP	OCT	NOV	DEC
Q												
W												
M												

Broken-belted Bumblebee

Bombus soroeensis

Habitat and distribution
Found in a variety of habitats, including late-flowering chalk downland, upland heaths and moors, woodland edges and coastal regions. Widespread; more frequent at higher altitudes in the Scottish Highlands, but a declining species further south. Absent from Ireland.

Mainly associated with Scottish moorlands, this rare bumblebee resembles a small version of the White-tailed Bumblebee aggregate, the workers often being particularly tiny. Its name refers to the partially interrupted yellow abdominal band, although this feature is not always easy to see and is not reliable for identification. Faded workers can be difficult to separate from those of similar species but the fluffy males are easier to identify, although the colour of the tail varies, particularly in northern populations, where it may be quite orange at its base. Compared to other bumblebees, queens appear late in the year, in May and June, and males between late July and November. Although capable of withstanding low temperatures at altitude, this species is also found sparingly further south, most notably on the large, often exposed expanses of flower-rich habitats of Salisbury Plain.

Worker on Devil's-bit Scabious

Nest

Nests are small to medium in size, with up to 150 workers. They are made under cover, usually underground, and often in old rodent nests.

Flowers visited

The species has a short tongue and visits clovers, White Dead-nettle, Common Bird's-foot Trefoil and, later in the year, heathers, scabiouses, legumes, knapweeds and Bramble.

Cuckoo parasites

None recorded.

Similar species

White-tailed Bumblebee aggregate (pages 26–27), Buff-tailed Bumblebee (pages 30–31), Early Bumblebee (pages 46–47), Southern Cuckoo Bumblebee (pages 80–81).

face

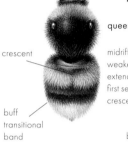

worker resembles queen, but often very small

queen

crescent

midriff band usually weakened in centre and extends onto sides of first segment, forming a crescent shape

buff transitional band

buff transitional band

typical male

buff-tailed male

fluffy, with weak midriff band

male hind leg very slender (cf. other similar bumblebees)

mandible lacks notch (cf. other similar bumblebees)

male genitalia

antenna segment 4 as wide as long (cf. other similar bumblebees)

	JAN	FEB	MAR	APR	MAY	JUN	JUL	AUG	SEP	OCT	NOV	DEC
Q												
W												
M												

Garden Bumblebee

Bombus hortorum

Habitat and distribution

Found in a wide range of habitats, from urban gardens, meadows and woodlands to mountainous regions. Widespread and common throughout Britain and Ireland, as far north as Shetland, but not usually seen in large numbers.

This distinctive and frequent garden visitor has the longest face and tongue of all our bumblebees, allowing it to forage on many long-tubed flowers, including Honeysuckle, Foxglove and Comfrey. It resembles other large common white-tailed species but is notable for having two central yellow bands, one on the scutellum at the rear of the thorax and the other at the base of the abdomen. These can often appear as one broad band, especially in flight. The species also has a shaggier coat than most other bumblebees and all castes have the same colour combinations, although darker (almost black) individuals occasionally occur. It most closely resembles the Ruderal Bumblebee, which also visits similar flower-rich habitats, although much less frequently. The Garden Bumblebee is one of the 'big seven' common garden species (see page 20).

Queen about to use her long tongue to probe into Foxglove

Nest

Nests are small, with up to 100 workers. They are usually made below ground in an old rodent nest but are sometimes on the surface among leaf litter. They have also occasionally been found above ground in old birds' nests. A second generation is sometimes produced.

Flowers visited

The species has a long tongue and visits a wide range of flowers. In spring these include clovers, dead-nettles, bluebells and Ground-ivy, and later in the year, Marjoram, Honeysuckle, thistles and vetches.

Cuckoo parasites

Barbut's Cuckoo Bumblebee (pages 70–71).

Similar species

See pages 36–37.

face

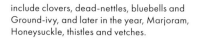

typical male
long, shaggy pile

queen   male

double yellow midriff band

dark male

upper fringe of male tibia longer than
width of tibia and uniform in length

male genitalia

Garden Bumblebee   Ruderal Bumblebee
                          Bombus ruderatus

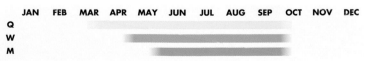

	JAN	FEB	MAR	APR	MAY	JUN	JUL	AUG	SEP	OCT	NOV	DEC
Q												
W												
M												

Garden Bumblebee: similar species

Below are illustrated differences between the Garden Bumblebee and other similar species with a double yellow midriff band and white tail.

Females

Garden Bumblebee
Bombus hortorum

usually larger and darker, with a neater, less shaggy body pile

Ruderal Bumblebee
Bombus ruderatus (pages 38–39)

smaller, with buff-yellow bands and a long body pile

face much shorter

mainly found on heaths and moors

Heath Bumblebee
Bombus jonellus (pages 42–43)

buff-yellow collar; midriff band weak

body hair sparse, giving a shiny appearance

outer face of hind tibia densely haired

Barbut's Cuckoo Bumblebee
Bombus barbutellus (pages 70–71)

Males

black face;
long, shaggy pile

yellow face

Garden Bumblebee
Bombus hortorum

White-tailed Bumblebee
Bombus lucorum
(pages 26–27)

neat, less shaggy pile;
yellow bands more
clearly defined

pale hairs spreading
into black

Ruderal Bumblebee
Bombus ruderatus
(pages 38–39)

smaller and more rounded, with a
long body pile

buff-yellow bands extensive,
extending onto face and lower
thorax

Heath Bumblebee
Bombus jonellus
(pages 42–43)

face short

buff-yellow collar

body hair fluffy and
sparse, giving a shiny
appearance

outer surface of hind tibia
densely haired

Barbut's Cuckoo Bumblebee
Bombus barbutellus
(pages 70–71)

37

Ruderal Bumblebee

Bombus ruderatus

Habitat and distribution

Found in farmland and occasionally in gardens, as well as on calcareous grassland and damper wetland areas. Its distribution has contracted, mainly to the East Midlands and south-east England, but there has been a slight increase in records in recent years. Absent from Ireland.

A large, robust-looking bumblebee, this species has a neat, short-coated appearance that may help distinguish it from the shaggier coat of the Garden Bumblebee. However, the two species can be very similar in appearance and sometimes it is almost impossible to separate them. Both have very long faces and tongues, well adapted for probing long-tubed flowers. The markings of both species vary, with some specimens (particularly Ruderal Bumblebee males) often being quite dark or even completely black. Ruderal Bumblebee queens usually appear from hibernation slightly later in the year, in April, and produce only a single generation that lasts into October. Towards the end of the 20th century the Ruderal Bumblebee became quite rare, but in recent years it has undergone a slight recovery.

Queen feeding on Red Clover, one of the species' favourite nectar sources

Nest

Nests are underground, often in an old rodent burrow, and are accessed by a sloping tunnel. They may contain up to 150 workers.

Flowers visited

A long-tongued species, the Ruderal Bumblebee visits White Dead-nettle, Red Clover, legumes, Viper's Bugloss and, in damper places, Comfrey and Marsh Woundwort.

Cuckoo parasites

None recorded.

Similar species

Garden Bumblebee (pages 34–35), Heath Bumblebee (pages 42–43), Short-haired Bumblebee (pages 52–53), Barbut's Cuckoo Bumblebee (pages 70–71).

body pile short and neat

band on tergite 1 usually faint

typical queen

melanic queen (form *perniger*)

face

upper fringe of male tibia shorter than width of tibia and of unequal length

typical male

body pile neat

male genitalia identical to Garden Bumblebee *B. hortorum*

Ruderal Bumblebee
Bombus ruderarius

Garden Bumblebee
Bombus hortorum

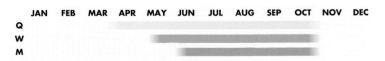

	JAN	FEB	MAR	APR	MAY	JUN	JUL	AUG	SEP	OCT	NOV	DEC
Q												
W												
M												

Tree Bumblebee

Bombus hypnorum

Mainly woodland, urban and suburban areas. Widespread and common throughout England and Wales, and has now expanded into Scotland and Ireland.

First recorded in Wiltshire in 2001, the Tree Bumblebee has become one of the most familiar bumblebees in gardens in England and Wales over the last decade, and has spread at a tremendous rate into Scotland and, more recently, Ireland. It is the most recent member of Britain's most common 'big seven' garden bumblebees, although its sudden appearance doesn't appear to have negatively affected other bumblebee species.

The Tree Bumblebee's distinctive ginger, black and white colouring makes it easily identifiable, but these markings vary and some specimens may be virtually all black apart from a white tail. Queens often hibernate in dead wood and emerge in February or March. Nests are made above ground, often in close proximity to one another, with workers tending to forage closer to their nests than those of most other species. New queens emerge in May and June, attracting swarms of males that sometimes clash with workers; this can cause concern if the nest is close to human habitation.

Queen foraging on a Bramble flower

Nest

Nests are usually well above ground, under the eaves of houses, in old rot holes in trees, in birds' nest holes and in nest boxes, but they are occasionally below ground. They may contain more than 150 workers. The species often produces a partial second generation.

Flowers visited

Short-faced and short-tongued, the Tree Bumblebee visits many cultivated and wild plants, including Raspberry, cotoneasters, rhododendrons, ceanothus, Common Snowberry, grape hyacinths, Bramble, Blackthorn and thistles. It also engages in robbing nectar from long-tubed flowers such as Comfrey.

Cuckoo parasites

None recorded.

Similar species

Common Carder Bee dark form (pages 62–63).

dark sides to thorax (cf. dark form of Common Carder Bee)

ginger
black
white

worker similar to queen, but smaller

typical queen

face

ginger hairs on face

typical male

male with ginger extending onto abdomen

queen dark form always with white tail

male mid-basitarsi lacking sharp point (cf. carder bees, page 61)

female hind basitarsi lacking sharp point (cf. carder bees, page 61)

male genitalia

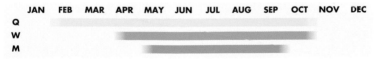

	JAN	FEB	MAR	APR	MAY	JUN	JUL	AUG	SEP	OCT	NOV	DEC
Q												
W												
M												

Heath Bumblebee

Bombus jonellus

This fairly small, very active bumblebee is easily overlooked. While it is typically found on heaths and moors in upland Scotland, where it can be quite common, it also occurs in a variety of flower-rich habitats in scattered populations throughout England, Wales and Ireland. Queens emerge in February or March in the south and later further north. They seek out nest sites either on or below ground, or in the roofs of houses or bird nest boxes.

The Heath Bumblebee resembles the Garden Bumblebee but is generally smaller, with buff-yellow rather than bright yellow bands. It also has longer body pile, a much rounder face and a short tongue, suitable for feeding from heathers, thistles, ragworts and umbellifers. Males and females have similar markings, but some Scottish island populations have a tail that is buff or orange rather than white.

Queen on Wild Thyme, showing her very short face and orange hairs fringing the pollen basket

Nest

Various nest sites are used, either on or below ground, among vegetation, or above ground in old birds' nests or buildings. Colonies are small, with around 50 workers, and short-lived. Two generations are produced in a season.

Flowers visited

This short-tongued species visits a wide range of flowers, including willows, legumes, clovers, Wild Thyme and heathers.

Cuckoo parasites

Forest Cuckoo Bumblebee (pages 78–79).

Similar species

Garden Bumblebee (pages 34–35), Ruderal Bumblebee (pages 38–39), Barbut's Cuckoo Bumblebee (pages 70–71).

face

long body pile; buff-yellow bands may be reduced or absent

worker similar to queen, but smaller

females from Scottish offshore islands have orange-buff tails

typical queen

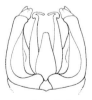

long, yellow hairs on face

male

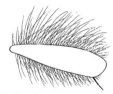

rotund with long body pile, long antennae and extensive yellow markings

male genitalia

male hind leg tibia
hairs much longer than width of tibia (cf. Short-haired Bumblebee B. subterraneus, page 53)

	JAN	FEB	MAR	APR	MAY	JUN	JUL	AUG	SEP	OCT	NOV	DEC
Q												
W												
M												

Bilberry Bumblebee

Bombus monticola

Mainly occurs on upland moorland and acidic grassland in northern, western and south-western Britain and parts of Ireland, but may also occur at sea level.

Uncommon and declining, this species is found mainly on upland moors and heaths in the north and west of Britain where there is an abundance of Bilberry. It prefers a mix of moorland and grassland where workers can forage on a variety of flowers throughout the year. Often found in the company of the Heath Bumblebee, it may occur at altitudes above 1,000m, where it is able to withstand low summer temperatures, and also at sea level.

Although quite small, this is a beautiful bumblebee with an extensive fox-red tail. Queens and workers are notably rounded in appearance, whereas the male, although similarly marked, has much longer hairs and a more conspicuous yellow head and face. Darker specimens occasionally occur and pale, faded specimens may appear later in the year.

Queen feeding on Bilberry

Nest

Nests are on or just below the ground in old rodent burrows. Colonies are small, with usually fewer than 100 workers.

Flowers visited

This fairly short-tongued species forages on willows, Bilberry, gorses and White Clover in spring, and heathers, Wild Thyme, Devil's-bit Scabious and rhododendrons later in the year. It also robs nectar from Bell Heather flowers.

Cuckoo parasites

Possibly the Forest Cuckoo Bumblebee (pages 78–79).

Similar species

Early Bumblebee (pages 46–47), Red-tailed Bumblebee (pages 54–55), Red-shanked Carder Bee (pages 66–67), Red-tailed Cuckoo Bumblebee (pages 76–77).

red hairs from tergite 2 to tip of tail

typical queen

dark worker

face

all castes have compact, rotund form

yellow hairs on face and head

male

male genitalia, similar to Early Bumblebee B. pratorum

female hind leg basitarsi lacking sharp point (cf. extinct Apple Bumblebee B. pomorum, page 85)

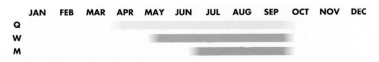

	JAN	FEB	MAR	APR	MAY	JUN	JUL	AUG	SEP	OCT	NOV	DEC
Q												
W												
M												

Early Bumblebee

Bombus pratorum

Widely distributed and found in many habitats, including gardens, woodland, moorland and coastal regions, but avoids more exposed habitats.

As its common name suggests, this active little bee is one of the earliest bumblebees to emerge from hibernation, in March or even February in mild weather. In the south, two generations appear, with individuals continuing into August and occasionally even later. One of the 'big seven' common garden bumblebees (see page 21), it is frequently seen in gardens and many other habitats throughout the British Isles.

The species is our smallest bumblebee, with typical specimens readily identified by the small, rusty-red tail, although there is some variation in the banding, which may be virtually absent in some workers. Later in the year the red tail often fades to whitish with age. Males are fluffy and rotund, with a characteristic yellow face. They appear in April or May, when they can be seen foraging busily in gardens from flowers such as lavenders, cotoneasters, *Pyracantha* and snowberries. The Early Bumblebee is also a notable pollinator of soft fruit in gardens and allotments.

Male feeding from *Eryngium*, showing his fluffy, dumpy body and yellow head

Nest

Nests are either above or below ground, and holes in trees or bird nest boxes may also be used. Colonies are short-lived and quite small, with up to 100 workers. Two generations are often produced in the south.

Flowers visited

This short-tongued species visits a wide range of plants, including, willows, Raspberry, Bramble, clovers, dead-nettles, thistles and knapweeds. It also engages in robbing nectar from long-tubed flowers such as Comfrey.

Cuckoo parasites

Forest Cuckoo Bumblebee (pages 78–79).

Similar species

See pages 48–49.

fluffy pile

dark worker

face

typical queen

typical worker resembles queen, but may be tiny with reduced markings

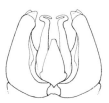

typical worker

rotund with fluffy pile

face and top of head yellow

male genitalia similar to Bilberry Bumblebee B. monticola

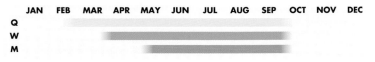

	JAN	FEB	MAR	APR	MAY	JUN	JUL	AUG	SEP	OCT	NOV	DEC
Q												
W												
M												

Early Bumblebee: similar species

Below are illustrated differences between the Early Bumblebee and other similar species with a yellow collar and midriff band, and a red tail.

Females

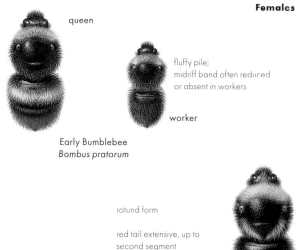

queen

fluffy pile;
midriff band often reduced
or absent in workers

worker

Early Bumblebee
Bombus pratorum

rotund form

red tail extensive, up to
second segment

mainly found on upland
moors and heaths

Bilberry Bumblebee
Bombus monticola
(pages 44–45)

typical females have white tails, but
those on the Scottish offshore islands
have orange-buff tails

Heath Bumblebee
Bombus jonellus
(pages 42–43)

fluffy and rotund

face and head
yellow

Early Bumblebee
Bombus pratorum

more elongate,
with a black face

yellow midriff band
sparsely haired

tail white to orange

Broken-belted
Bumblebee
Bombus soroensis
(pages 32–33)

longer and more
elongated

yellow face
and head

weak midriff band

may become
sun-bleached with age

Red-tailed Bumblebee
Bombus lapidarius
(pages 54–55)

fluffy and rotund;
face and top of head yellow;
red tail extensive

mainly found on upland
moors and heaths

Bilberry Bumblebee
Bombus monticola
(pages 44–45)

collar and midriff
bands duller,
olive-grey

Red-shanked Carder
Bee (pale form)
Bombus ruderarius
(pages 66–67)

some old and faded
specimens of bumblebees
with red tails may resemble
this species

Shrill Carder Bee
Bombus sylvarum
(pages 68–69)

Great Yellow Bumblebee

Bombus distinguendus

Habitat and distribution

Exposed flower-rich coastal machair grassland and dunes, along the north coast of mainland Scotland, Orkney and some islands of the Hebrides, and western Ireland.

This spectacular large bumblebee was once sparsely widespread throughout Britain and Ireland, but its numbers have dwindled in the last 100 years to just five population centres. It is now one of our two bumblebees regarded as nationally endangered, the other being the Shrill Carder Bee.

Queens emerge from hibernation relatively late in the year, from mid-May to early June, and produce workers in July and males in August. With their bright mustard-yellow colour, both workers and males resemble the queen, and apart from the width of the black thoracic band, there is little variation. The male can be virtually identical to the closely related Short-haired Bumblebee, but as the distributions of both have contracted drastically, there is little chance of confusion in identification. Another rare bee, the Moss Carder Bee (pages 60–61), with which the Great Yellow Bumblebee sometimes flies, lacks the black band on the thorax.

Queen on Common Bird's-foot Trefoil

Nest

Nests are sited beneath grassy tussocks or below ground, often in old rodent holes. They are small, with sometimes fewer than 20 workers.

Flowers visited

With its fairly long tongue, the Great Yellow Bumblebee visits Kidney Vetch, Common Bird's-foot Trefoil, clovers (particularly Red Clover), Common Knapweed, thistles, Devil's-bit Scabious and Common Ragwort.

Cuckoo parasites

None recorded.

Similar species

Short-haired Bumblebee male (pages 52–53), Field Cuckoo Bumblebee male (pages 74–75), carder bees (pages 58–69).

neat, short pile

black thoracic band varies in width

yellow may become sun-bleached with age

face

queen

worker similar to queen, but smaller and with longer body pile

longer pile than queen, but similarly marked

male

male gonitalia

	JAN	FEB	MAR	APR	MAY	JUN	JUL	AUG	SEP	OCT	NOV	DEC
Q												
W												
M												

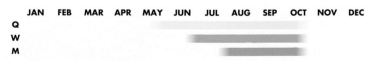

51

Short-haired Bumblebee

Bombus subterraneus

Habitat and distribution

Favours large, open areas of flower-rich grasslands, shingle and downland. Formerly widespread, the species is now extinct and occurs only at a single reintroduction site in Kent.

Once widespread throughout England as far north as Yorkshire, and particularly common in Kent and Suffolk, the Short-haired Bumblebee suffered serious decline throughout the second half of the 20th century and became extinct in Britain in the 1980s. Its decline was inevitably connected to the destruction of suitable large, open, flower-rich grasslands; joint conservation projects in south-east England are now in the process of re-creating this habitat. Earlier this century, queens taken from stock introduced to New Zealand as pollinators of Red Clover in the 19th century were brought back to the UK in an attempt at reintroduction. However, this was unsuccessful, as was a more recent attempt to introduce bees from Sweden.

Queens most closely resemble those of the Ruderal Bumblebee, but have much shorter hair and a slightly shorter face. Males can be very similar to the Great Yellow Bumblebee, but the distributions of the two species do not overlap.

Worker feeding from Red Clover

52

Nest
Nests are sited underground in old rodent holes and are said to have an unpleasant odour. They are of medium size, with 75–100 workers.

Flowers visited
This long-tongued species visits the flowers of legumes such as Red Clover, along with Viper's Bugloss, Bramble, Honeysuckle and White Dead-nettle.

Cuckoo parasites
None recorded.

Similar species
Ruderal Bumblebee (pages 38–39), Great Yellow Bumblebee male (pages 50–51), Field Cuckoo Bumblebee male (pages 74–75).

pile shorter and neater than in any other bumblebee

may resemble Ruderal Bumblebee *B. ruderatus* (pages 38–39)

tail dirty white

queen

face

pale specimens may resemble male Field Cuckoo Bumblebee *B. campestris* (pages 74–75)

male

central keel on sternite 6 of female

male hind tibia with short dorsal fringe

male genitalia similar to Great Yellow Bumblebee

Short-haired Bumblebee

Ruderal Bumblebee *Bombus ruderatus* (pages 38–39)

Red-tailed Bumblebee

Bombus lapidarius

Queens of this distinctive large bumblebee often hibernate communally and are frequently seen in gardens and urban areas from mid-February, seeking out early-flowering plants such as Dandelion, willows and gorses. The species is common and widely distributed throughout Britain and Ireland, and is one of the 'big seven' common garden bumblebees.

Workers resemble the queen, which has a glossy black coat when freshly emerged, but they can be very small. The male retains the red tail but has extensive yellow hairs on his face, head and thorax, although this characteristic is variable. Workers and males can fade with age, often causing some difficulty with identification. The Red-shanked Carder Bee is most likely to cause confusion, but it is smaller, more rotund and much rarer.

Habitat and distribution

Found in many flower-rich habitats but tends to avoid high, exposed regions. Widespread; more common in the south but gradually spreading further north.

Worker feeding from Purple Top, *Verbena bonariensis*, a good nectar source in many gardens

Nest

Nests are large, often with more than 150 workers, and are usually underground in a variety of locations, often near walls, bare ground or beneath rocks.

Flowers visited

A short-tongued species, the Red-tailed Bumblebee visits a wide range of cultivated plants and wildflowers, especially composites such as Dandelion, asters and thistles, as well as willows, Common Gorse, and many crucifers and legumes.

Cuckoo parasites

Red-tailed Cuckoo Bumblebee (pages 76–77).

Similar species

See pages 56–57.

large, glossy and jet black, with a crimson-red tail

extensive yellow hairs on head and collar

face

worker similar, but smaller and fluffier

queen

male

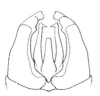

female hind leg with black hair fringes (cf. Red-shanked Carder Bee *B. ruderarius*, page 67)

faded sun-bleached male

male genitalia

Red-tailed Bumblebee

Cullum's Bumblebee *B. cullumanus* (page 84)

	JAN	FEB	MAR	APR	MAY	JUN	JUL	AUG	SEP	OCT	NOV	DEC
Q												
W												
M												

55

Red-tailed Bumblebee: similar species

Below are illustrated differences between the Red-tailed Bumblebee and other similar species that are predominantly black with a red tail.

Females

large and distinctive, with a crimson-red tail

hind tibia shiny and fringed with black hairs

Red-tailed Bumblebee
Bombus lapidarius

smaller and more rotund, with orange-red tail

hind tibia shiny and fringed with orange hairs

Red-shanked Carder Bee
Bombus ruderarius (pages 66–67)

large and shiny, with sparse body hair

head very large and box-shaped; wings dark

hind tibia densely haired

Red-tailed Cuckoo Bumblebee
Bombus rupestris (pages 76–77)

very different from female

extensive yellow hairs on head, collar and midriff

Red-tailed Bumblebee
Bombus lapidarius

collar and midriff bands are dull olive-grey

Red-shanked Carder Bee
Bombus ruderarius
(pages 66–67)

rotund and fluffy

bright yellow collar and midriff

Early Bumblebee
Bombus pratorum
(pages 46–47)

rotund and fluffy

extensive red tail

mainly on upland heaths and moors

Bilberry Bumblebee
Bombus monticola (pages 44–45)

olive bands are often absent

head large and box-shaped; body shiny

Red-tailed Cuckoo Bumblebee
Bombus rupestris
(pages 76–77)

Brown-banded Carder Bee

Bombus humilis

One of three similar species of ginger carder bees and one of the rarest bumblebees in Britain, the Brown-banded Carder Bee has seen a drastic contraction in its range over the last 70 years, due mainly to agricultural intensification and the loss of flower-rich grasslands. However, in recent years, the species' range has expanded again.

Despite its common name, the brown band across the abdomen is not always obvious and in some specimens it may be necessary to examine other features, such as the male genitalia, for positive identification. As with all the carder bees, nests are covered in moss, and grass is raked or 'carded' over the top by the queen and workers. Foraging usually occurs within a kilometre of the nest site. The similar and very rare Moss Carder Bee has a more northerly distribution, but the two species sometimes occur together, which can cause confusion over identification.

Habitat and distribution

Warm, open, well-vegetated places, including downland, coastal dunes and brownfield sites. Formerly widespread but its range has contracted, mainly to western and southern coastal regions. Other strongholds include the Thames Estuary and Salisbury Plain. There have been recent records from some Midland counties. Absent from Ireland.

Worker on Tufted Vetch

Nest
Nests are found in tall, open grassland, among leaf litter on the surface of the ground, and carded in moss and grass. They are small, usually with around 50 workers.

Flowers visited
This moderately long-tongued species is attracted to vetches, trefoils, clovers, thistles, knapweeds, scabiouses and Honeysuckle.

Cuckoo parasites
Field Cuckoo Bumblebee (pages 74–75).

Similar species
Moss Carder Bee (pages 60–61), Common Carder Bee (pages 62–63).

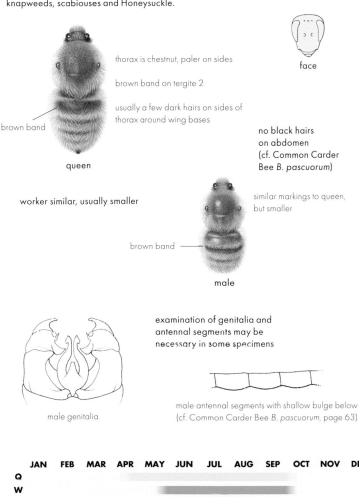

thorax is chestnut, paler on sides

brown band on tergite 2

usually a few dark hairs on sides of thorax around wing bases

brown band

queen

face

no black hairs on abdomen (cf. Common Carder Bee *B. pascuorum*)

worker similar, usually smaller

similar markings to queen, but smaller

brown band

male

examination of genitalia and antennal segments may be necessary in some specimens

male genitalia

male antennal segments with shallow bulge below (cf. Common Carder Bee *B. pascuorum*, page 63)

	JAN	FEB	MAR	APR	MAY	JUN	JUL	AUG	SEP	OCT	NOV	DEC
Q												
W												
M												

Moss Carder Bee

Bombus muscorum

Habitat and distribution

Prefers cool, damp grasslands, marshes, dunes and sea banks. Once widespread, the species is still fairly common on coastal and moorland sites in northern Britain, but very local in England and Wales. Widespread in Ireland and also recorded from the Channel Islands.

This is the largest of the three ginger carder bees and, like the similar Brown-banded Carder Bee, is one of the rarest and fastest-declining of our bumblebees, particularly in the south. Specimens from the mainland have a beautiful bright ginger thoracic pile, usually surrounded with a distinct pale creamy border. Unlike other ginger carder bees, they have no black hairs on the thorax; however, faded specimens of any of the three ginger carder species may be difficult to separate.

Some island populations in the far north of Scotland and Ireland are boldly marked with rust red and black, rivalling the markings of the Tree Bumblebee but without the white tail; several of these variants have been given their own names. Male Moss Carder Bees are similar to, but slightly larger than, the workers, which are said to be touchy and defensive if their nest is threatened.

Queen on Kidney Vetch

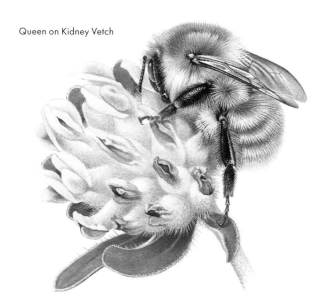

Nest

Nests are usually on or just below ground level, often in tall, open grassland, where they may be vulnerable to overgrazing and mowing. They usually contain fewer than 100 workers.

Flowers visited

The long-tongued Moss Carder Bee visits various clovers and vetches, Red Dead-nettle, thistles, heathers, teasels, scabiouses and Red Bartsia.

Cuckoo parasites

Field Cuckoo Bumblebee (pages 74–75).

Similar species

Tree Bumblebee (pages 40–41), Brown-banded Carder Bee (pages 58–59), Common Carder Bee (pages 62–63).

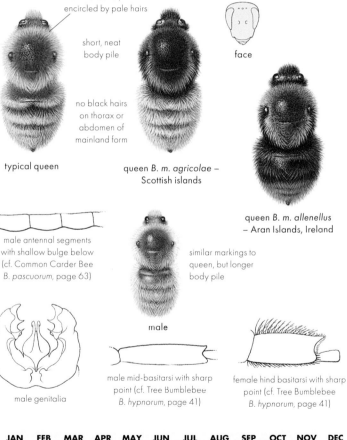

thorax chestnut, encircled by pale hairs

short, neat body pile

no black hairs on thorax or abdomen of mainland form

typical queen

queen *B. m. agricolae* – Scottish islands

face

queen *B. m. allenellus* – Aran Islands, Ireland

male antennal segments with shallow bulge below (cf. Common Carder Bee *B. pascuorum*, page 63)

similar markings to queen, but longer body pile

male

male genitalia

male mid-basitarsi with sharp point (cf. Tree Bumblebee *B. hypnorum*, page 41)

female hind basitarsi with sharp point (cf. Tree Bumblebee *B. hypnorum*, page 41)

	JAN	FEB	MAR	APR	MAY	JUN	JUL	AUG	SEP	OCT	NOV	DEC
Q												
W												
M												

Common Carder Bee

Bombus pascuorum

Habitat and distribution
Ubiquitous, including gardens and urban sites. Common and widespread throughout the British Isles.

Probably our most common and widespread bumblebee, the Common Carder Bee is found in many habitats throughout Britain and Ireland, including gardens, and is one of the 'big seven' regularly observed garden bumblebees. It is a very variable species, with northern bees often much paler than those in the south. Workers and males resemble the queen, but the former may be very small, and all castes are noticeably fluffier than those of the other two neater and rarer ginger carder bees.

The queen is the first of the carder bees to emerge from hibernation in March, when she seeks out early-flowering plants such as White Deadnettle and Dandelion. The flight period is longer than that of any other bumblebee, continuing into November in two generations. Like all other carder bees, the queen rakes grass and moss over the nest to conceal and protect it.

Queen on a Pea flower – the species is an important pollinator of crops and fruit trees

Nest
Nests are usually on or just below the ground, often in long grass or among vegetation, but occasionally well above ground. They contain 60–200 workers.

Flowers visited
This long-tongued species visits a huge range of plants and is well adapted to foraging among those with deep flowers, such as dead-nettles, Comfrey and Foxglove.

Cuckoo parasite
Field Cuckoo Bumblebee (pages 74–75).

Similar species
See pages 64–65.

fairly long and uneven body pile

thorax uniform but variable in colour

abdomen with varying amounts of black

worker similar, but often very small

typical queen

face

different colour forms may occur together

pale queen

black hairs on abdomen often form bands

typical male

pale male

dark male

male genitalia

examination of genitalia and antennal segments may be necessary in some specimens

male antennal segments with pronounced bulge below (cf. Brown-banded and Moss Carder Bees B. humilis and B. muscorum, pages 59 and 61)

	JAN	FEB	MAR	APR	MAY	JUN	JUL	AUG	SEP	OCT	NOV	DEC
Q												
W												
M												

Common Carder Bee: similar species

Below are illustrated differences between the Common Carder Bee and other similar species with a buff thorax.

Females

body pile long and uneven

thorax uniform in colour but variable

abdomen usually with black hairs, sometimes pale

Common Carder Bee
Bombus pascuorum

thorax chestnut, paler on sides

brown band on tergite 2

usually a few dark hairs on sides of thorax around wing bases

Brown-banded Carder Bee
Bombus humilis
(pages 58–59)

larger

pile neat

thorax chestnut, encircled by pale hairs

no black hairs on thorax and abdomen

Workers of all three species resemble small queens

Moss Carder Bee
Bombus muscorum (pages 60–61)

Moss Carder Bee
*Bombus muscorum
allenellus* – Aran Islands, Ireland (page 61)

dark sides

always with white tail

Tree Bumblebee
Bombus hypnorum
(pages 40–41)

usually with black hairs on abdomen and orange tail, but some specimens may be quite pale

Common Carder Bee
Bombus pascuorum

sides of thorax pale cream

brown band on tergite 2

Brown-banded Carder Bee
Bombus humilis (pages 58–59)

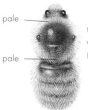

pale

pale

thorax chestnut, encircled by pale hairs

examination of genitalia and antennal segments may be necessary in some specimens

Moss Carder Bee
Bombus muscorum
(pages 60–61)

always has a white tail

Tree Bumblebee
Bombus hypnorum
(pages 40–41)

65

Red-shanked Carder Bee

Bombus ruderarius

Habitat and distribution

Occurs in open flower-rich habitats, including urban brownfield sites. Once widespread, it has declined and now occurs mainly in central, southern and south-east England, with isolated populations in the Inner Hebrides and the Channel Islands. Rare in western coastal parts of Ireland.

Queens and workers of this rare carder bee bear a close resemblance to the much more common Red-tailed Bumblebee but are generally smaller and more rotund, with a rather more orange tail; there is little variation. They also have a fringe of orange hairs around their pollen baskets, hence the common name. The male tends to be rather dull, although there is some variation, and he has a black-haired face, unlike the yellow-haired face of the male Red-tailed Bumblebee. However, with some males it may be necessary to examine the genitalia to separate them from other very similar species.

Queens are quite late to emerge from hibernation, in April, and the short-lived colonies last until late August. The Red-shanked Carder Bee suffered serious declines towards the end of the last century, causing great concern to conservationists.

Worker on Garden Lavender

Nest

Nests are sited among rank vegetation on the ground, often in old rodent nests, and are covered in shredded moss and grass. They usually contain fewer than 100 workers.

Flowers visited

A fairly long-tongued bumblebee, the species visits various legumes, labiates, Bramble, Viper's Bugloss and thistles.

Cuckoo parasites

None recorded.

Similar species

Red-tailed Bumblebee (pages 54–55), Red-tailed Cuckoo Bumblebee (pages 76–77).

quite small, rotund and fluffy

tail orange-red

hind tibia shiny, fringed with orange hairs

queen

face

males

males vary, but usually dull olive-grey with a pale collar, head small

worker hind tibia with pollen load, showing orange hair fringes

male genitalia

male hind basitarsi short and broad, with short dorsal fringe (cf. Red-tailed Cuckoo Bumblebee B. rupestris, page 77)

	JAN	FEB	MAR	APR	MAY	JUN	JUL	AUG	SEP	OCT	NOV	DEC
Q												
W												
M												

Shrill Carder Bee

Bombus sylvarum

A lowland species of flower-rich meadows, arable and brownfield sites, dunes, coastal sea banks and shingle beaches. Formerly common and widespread, it is now restricted to around half a dozen or so populations, its main strongholds being the Thames Gateway region, Somerset, south Wales and western Ireland.

This small bumblebee is one of the last species to emerge from hibernation in May. Although it is widespread, in England and Wales it is the rarest and most endangered bumblebee species. All three castes resemble one another, being greyish green with a black band through the thorax, a banded abdomen and a pale orange tail. They differ only in size, with workers being particularly small. Fresh specimens may be confused with worn Early Bumblebees but they often also fade with age, appearing greyish white but still with a noticeable pinkish-orange tail.

Formerly known as the Knapweed Carder Bee, the species' current vernacular name refers to the high-pitched hum it makes. This resembles the hum of a honeybee, and is most audible in the queen during her distinctive hovering flight as she forages from flower to flower. For people with reasonable hearing, this can be a good way of detecting the species.

Queen on
Red Bartsia

68

Nest
Nests are sited on or just below ground, among tall vegetation, often in an old rodent burrow. They are small, with about 100 workers.

Flowers visited
A long-tongued species, the Shrill Carder Bee favours Red Clover and other legumes, White Dead-nettle, knapweeds, Red Bartsia, thistles and Common Ragwort.

Cuckoo parasites
None recorded.

Similar species
Bilberry Bumblebee male (pages 44–45), Early Bumblebee male (pages 46–47), Red-tailed Bumblebee male (pages 54–55).

generally greyish green with a central black thorax

tail rust-tinted

face

queen

worker similar to queen, but smaller

worker

The dark band through the thorax and the orange tail help separate the Shrill Carder Bee from similar buff-coloured or faded bumblebees

male similar to female, but often with a brighter tail

male

male genitalia

	JAN	FEB	MAR	APR	MAY	JUN	JUL	AUG	SEP	OCT	NOV	DEC
Q												
W												
M												

Barbut's Cuckoo Bumblebee

Bombus barbutellus

Habitat and distribution
Like is hosts, Barbut's Cuckoo Bumblebee occurs in a wide range of flowery habitats. It is widespread but rarely frequent, and declining, found throughout much of central southern England but much rarer in the north and west.

Like several other cuckoo bumblebees, Barbut's Cuckoo Bumblebee closely resembles its host species, and because of this it is quite likely to have been under-recorded. However, its shiny abdomen and the absence of pollen baskets on the hind legs of the female indicate that it is not a true bumblebee. It also differs from true bumblebees in having a shorter, more bulky head and a short tongue.

Females emerge from hibernation in mid-April, a few weeks after queens of the host species, and the scruffy-haired males appear initially in June and continue until September. If there is any doubt about identification, both sexes can be quite easily identified by examining the tips of the underside of their abdomens (see opposite). The species is sometimes found in gardens, where its most common host is the Garden Bumblebee.

Male feeding on Red Clover

Flowers visited

This short-tongued species visits and rests on a range of flowers, including dead-nettles, vetches, thistles, knapweeds and Bramble.

Similar species

Garden Bumblebee (pages 34–35), Ruderal Bumblebee (pages 38–39), Heath Bumblebee (pages 42–43).

Hosts

Garden Bumblebee (pages 34–35), Ruderal Bumblebee (pages 38–39).

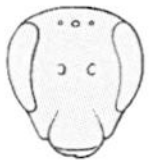

face

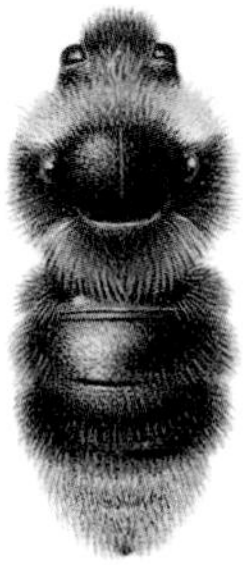

buff-yellow collar

midriff band weak

body hair sparse, giving a shiny appearance

male similar to female, but smaller and slimmer

white tail starts at tergite 4

female

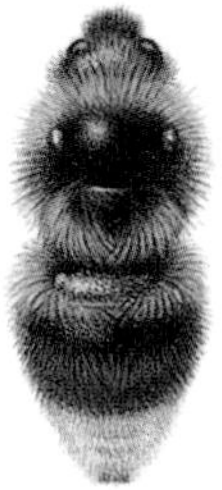

male

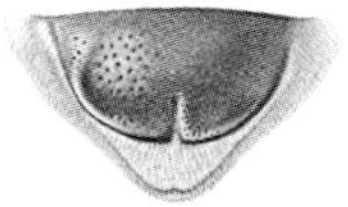

female sternite 6

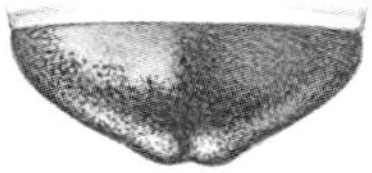

male sternite 6

examining the sternites is a good way to separate cuckoo bumblebees

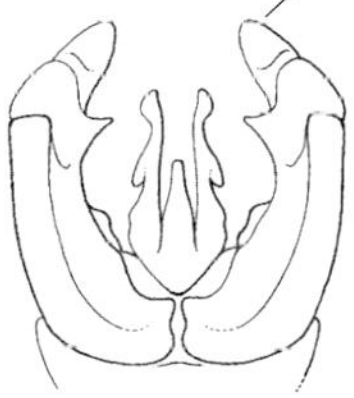

blunt tip

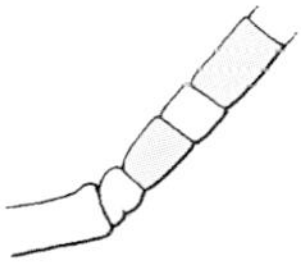

male antennal segment 3 shorter than segment 5

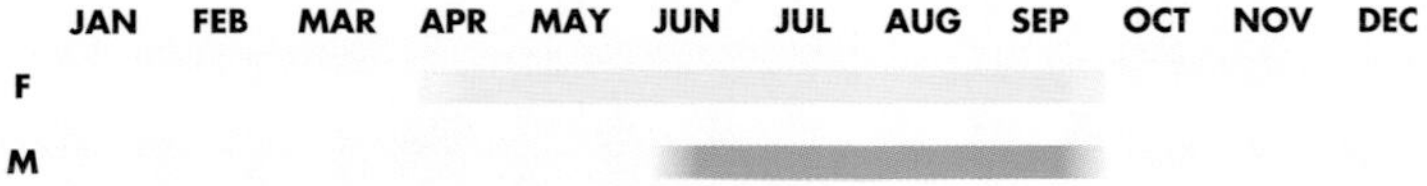

male genitalia

	JAN	FEB	MAR	APR	MAY	JUN	JUL	AUG	SEP	OCT	NOV	DEC
F												
M												

Gypsy Cuckoo Bumblebee

Bombus bohemicus

Habitat and distribution
Locally common in northern and western Britain, but scarcer and declining further south. Widespread in Ireland.

This cuckoo bumblebee is closely related to the very similar but slightly larger Southern Cuckoo Bumblebee, from which it can be differentiated by the paler, broader yellow hairs on the collar and the less intense black markings, giving it an overall paler appearance. The yellow hairs at the base of the white tail are also paler and less conspicuous. The male has a fluffy appearance, and although the genitalia of the two species are quite similar, their antennal segments differ and are perhaps the best way of separating them (see the illustrations here and on page 81).

The Gypsy Cuckoo Bumblebee is more likely to be seen in northern upland heaths, where it may be quite common; further south, there appears to be an overall general decline in populations. As with other cuckoo bumblebees, its wings are of a darker hue than those of true bumblebees.

Queen foraging on Bugle

Flowers visited

This short-tongued species visits a wide range of flowers, including Bramble, heathers, thistles and knapweeds, all for its own consumption.

Host

White-tailed Bumblebee aggregate (pages 26–27).

Similar species

White-tailed Bumblebee aggregate (pages 26–27), Southern Cuckoo Bumblebee (pages 80–81).

face

broad, pale yellow collar

body pile slightly fluffy but hairs sparse, giving a shiny appearance

pale yellow hairs on tergite 3

female

broad, pale yellow collar and midriff

body pile fluffy but sparse, giving a shiny appearance

pale yellow hairs on tergite 3

male

female sternite 6 very similar to Southern Cuckoo Bumblebee *B. vestalis*

male genitalia

male antennal segment 3 about equal to segment 5

	JAN	FEB	MAR	APR	MAY	JUN	JUL	AUG	SEP	OCT	NOV	DEC
F												
M												

Field Cuckoo Bumblebee

Bombus campestris

Occurs in a wide range of habitats, including urban areas, woodlands and flower-rich places. Widely distributed and fairly frequent in the south, but more local further north and rare in Scotland and Ireland.

The male of this widespread social parasitic bee can be quite variable in its markings and is one of the few bumblebees that often produce all-black specimens. The female is less variable and has a broad, buff band at the front of the thorax and a characteristic yellowish-buff tail with a central black divide. As with other cuckoo bumblebees, its abdomen is shiny black and the wings are dark-tinted.

The female first appears in April, although both females and males have been observed in the autumn, suggesting a possible second generation. This would correspond with the life cycle of the Common Carder Bee, which is the species' main host.

Pale male feeding on Devil's-bit Scabious

Flowers visited

The short-tongued Field Cuckoo Bumblebee favours White Clover, knapweeds, thistles, Devil's-bit Scabious and Bramble.

Hosts

Common Carder Bee (pages 62–63); probably also Brown-banded Carder Bee (pages 58–59) and Moss Carder Bee (pages 60–61).

Similar species

Great Yellow Bumblebee male (pages 50–51), Short-haired Bumblebee male (pages 52–53), Barbut's Cuckoo Bumblebee (pages 70–71).

one of three smaller cuckoo bumblebees

broad buff collar and weaker midriff

body pile sparse, giving a shiny appearance

buff tail divided by central black hairs

female

face

black bands

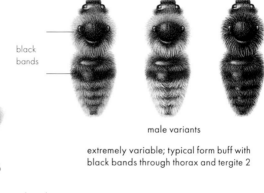

male variants

extremely variable; typical form buff with black bands through thorax and tergite 2

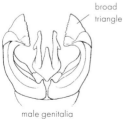

female sternite 6

broad triangle

male genitalia

male sternite 6 with lateral tufts

JAN	FEB	MAR	APR	MAY	JUN	JUL	AUG	SEP	OCT	NOV	DEC

F

M

Red-tailed Cuckoo Bumblebee

Bombus rupestris

This is the largest and most distinctive of the cuckoo bumblebees. The impressive female resembles her host, the Red-tailed Bumblebee, but her large head, shiny abdomen, very dark wings and slightly more orange-red tail make separation of the two species straightforward. The male, however, is more variable, most closely resembling the male Red-shanked Carder Bee. More detailed examination is sometimes needed to be certain of identification.

Males often congregate on hillsides in courtship leks. The female emerges from hibernation in May, and males and new females appear from mid-July to August. Although the species' host has always been common, its abundance fluctuated throughout the last century and declined in the latter half, but in the last couple of decades it has made a recovery and is sometimes quite frequent in the south of England.

Habitat and distribution

Found in a variety of flower-rich habitats wherever its host occurs. Widespread throughout Britain and Ireland with the exception of Scotland, and most frequent in southern and south-eastern England.

Large female feeding from knapweed, showing the species' broad box-shaped head

Flowers visited

The species has a medium-length tongue and visits White Clover, Dandelion, thistles, Bramble and knapweeds.

Similar species

Red-tailed Bumblebee (pages 54–55), Red-shanked Carder Bee (pages 66–67). (See also pages 56–57.)

Host

Red-tailed Bumblebee (pages 54–55).

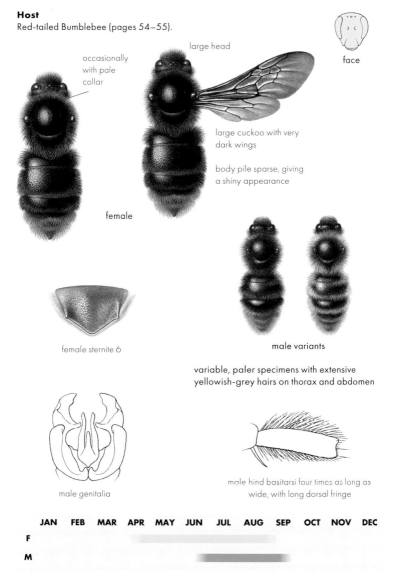

occasionally with pale collar

large head

face

female

large cuckoo with very dark wings

body pile sparse, giving a shiny appearance

female sternite 6

male variants

variable, paler specimens with extensive yellowish-grey hairs on thorax and abdomen

male genitalia

male hind basitarsi four times as long as wide, with long dorsal fringe

	JAN	FEB	MAR	APR	MAY	JUN	JUL	AUG	SEP	OCT	NOV	DEC
F												
M												

Forest Cuckoo Bumblebee

Bombus sylvestris

Habitat and distribution

Urban and rural areas, heathland and woodland. The species' habitat and distribution corresponds with that of its main host, the Early Bumblebee.

The female of this smallish bee is the first of the cuckoo bumblebees to emerge from hibernation, in late March, when its main host, the Early Bumblebee, is on the wing. The female is notable for the distinctive downcurved tip to her abdomen. The male also has a unique tail, which is fringed with orange hairs and has a black tip, although this characteristic may be difficult to see. This colour combination gave rise to the species' earlier name of Four-coloured Cuckoo Bumblebee.

Males appear in May and often congregate in woodland margins, giving off a distinctive musty odour. In southern Britain the Forest Cuckoo Bumblebee is thought to have two generations, when females may be seen as late as September.

Male feeding from Creeping Thistle

Flowers visited
This short-tongued bumblebee visits willows, Dandelion and dead-nettles, and, later, thistles, knapweeds and Devil's-bit Scabious.

Hosts
Early Bumblebee (pages 46–47); probably also Heath Bumblebee (pages 42–43) and Bilberry Bumblebee (pages 44–45).

Similar species
Field Cuckoo Bumblebee male (pages 74–75).

scutellum always dark-haired

face

female

small, fluffy cuckoo bee with a strongly downcurved tail

midriff band weak or absent

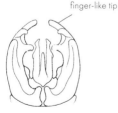

female sternite 6

scutellum dark-haired

male variants

variable

typical male is yellow-buff, black and white, with orange tail tip

melanic specimens often occur

finger-like tip

male genitalia

	JAN	FEB	MAR	APR	MAY	JUN	JUL	AUG	SEP	OCT	NOV	DEC
F												
M												

Southern Cuckoo Bumblebee

Bombus vestalis

Habitat and distribution

Found in similar habitats to its host, including flowery meadows, gardens, woodland, heathland and coasts. Widespread throughout England and parts of Wales, and more recently recorded from Scotland and Ireland.

This species is Britain's commonest cuckoo bumblebee, most frequent in the south but rare in Scotland and Ireland. Its distribution overlaps that of the similar Gypsy Cuckoo Bumblebee, which has its main stronghold in the Scottish Highlands. The female Southern Cuckoo Bumblebee is usually larger, with a darker ginger collar and brighter yellow patches at either side of the white tail. The male is similar to the female but smaller and with much longer antennae. He has a much neater appearance than the male Gypsy Cuckoo Bumblebee, but sometimes (especially in worn, sun-bleached specimens) close examination of the antennae may be required to separate the two species.

Males appear in May, and after mating can often be observed in gardens in late summer, feeding lazily from lavenders, thistles and chives. They continue into late August, when they die, leaving the females to enter hibernation.

Male feeding from knapweed, showing very long antennae

80

Flowers visited

A short-tongued species, the Southern Cuckoo Bumblebee visits a wide range of plants, including willows, *Prunus*, Ground-ivy, vetches, clovers, thistles, lavenders and hebes.

Host

Buff-tailed Bumblebee (pages 30–31).

Similar species

Gypsy Cuckoo Bumblebee (pages 72–73).

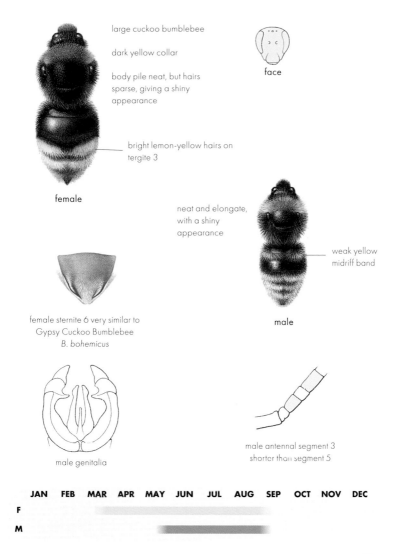

large cuckoo bumblebee

dark yellow collar

body pile neat, but hairs sparse, giving a shiny appearance

face

bright lemon-yellow hairs on tergite 3

female

neat and elongate, with a shiny appearance

weak yellow midriff band

female sternite 6 very similar to Gypsy Cuckoo Bumblebee *B. bohemicus*

male

male genitalia

male antennal segment 3 shorter than segment 5

	JAN	FEB	MAR	APR	MAY	JUN	JUL	AUG	SEP	OCT	NOV	DEC
F												
M												

Identifying male cuckoo bumblebees

Some male cuckoo bumblebees can resemble male true bumblebees, but the outer surface of the hind tibia of cuckoo bumblebees is densely hairy and, viewed from above, the head is broad and box-shaped. Male Forest Cuckoo Bumblebees are variable and can be confused with other species; if there is any doubt, the features highlighted below should help with identification, although microscopic examination may be necessary.

genitalia

antennal segments 3 and 5 of equal length

Gypsy Cuckoo Bumblebee
Bombus bohemicus (pages 72–73)

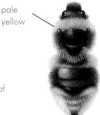

pale yellow

long body pile

genitalia

antennal segment 3 shorter than segment 5

Southern Cuckoo Bumblebee
Bombus vestalis (pages 80–81)

neat body pile

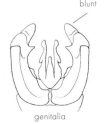

genitalia

blunt

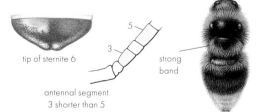

tip of sternite 6

antennal segment 3 shorter than 5

strong band

Barbut's Cuckoo Bumblebee
Bombus barbutellus (pages 70–71)

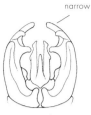
narrow

genitalia

black and orange tip to tail

Forest Cuckoo Bumblebee
Bombus sylvestris (pages 78–79)

broad and triangular

genitalia

hair tufts on sternite 6

variants

Field Cuckoo Bumblebee
Bombus campestris (pages 74–75)

broad, crab-like

genitalia

Red-tailed Cuckoo Bumblebee
Bombus rupestris (pages 76–77)

83

Cullum's Bumblebee

Bombus cullumanus

Cullum's Bumblebee is a true bumblebee that is presumed extinct in the British Isles. The queen closely resembles the Red-tailed Bumblebee (see pages 54–55), but is smaller and has a much shorter face (although workers of both species are about the same size). Some female specimens also have an obscure pale collar on the thorax, behind the head. The male is more distinctive, with a dark abdominal band and an orange tail and thoracic collar.

Cullum's Bumblebee was first recorded in Britain when a male was caught in Suffolk nearly 200 years ago, and it was last recorded on the Berkshire Downs (now Oxfordshire) in 1941. Always a rare species, it was confined to sheep walks and large tracts of flower-rich calcareous grassland in central and southern England.

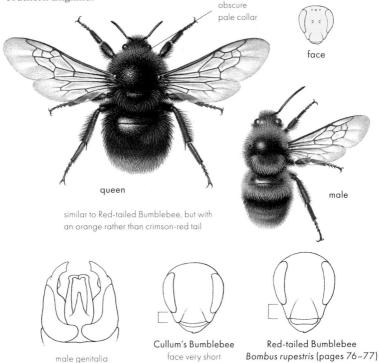

obscure pale collar

face

queen

similar to Red-tailed Bumblebee, but with an orange rather than crimson-red tail

male

male genitalia

Cullum's Bumblebee
face very short

Red-tailed Bumblebee
Bombus rupestris (pages 76–77)

84

Apple Bumblebee

Bombus pomorum

This true bumblebee species is presumed extinct in Britain, although some scientists think that it was never a permanent resident here. Only four British specimens are known, three males and a queen, all of which were found in Kent in the mid-19th century and are now kept at the Oxford University Museum of Natural History.

The queen resembles a small Red-tailed Bumblebee (see pages 54–55), but with pale bands on the thorax and a more extensively red abdomen, rather like that of the Bilberry Bumblebee (see pages 44–45). The male is similar to the queen but brighter, rather like a pale Red-tailed Cuckoo Bumblebee (see pages 76–77). This long-faced species is said to favour legume, labiate and composite flowers.

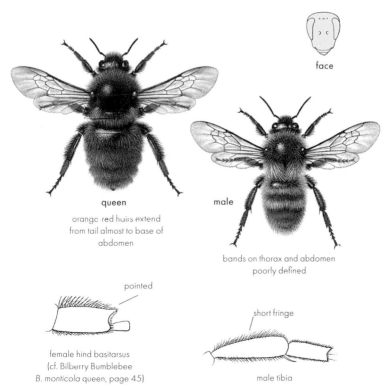

face

queen

orange-red hairs extend from tail almost to base of abdomen

male

bands on thorax and abdomen poorly defined

pointed

female hind basitarsus (cf. Bilberry Bumblebee *B. monticola* queen, page 45)

short fringe

male tibia

85

Mimics and lookalikes

Several species of insect gain protection from predators by resembling bees, wasps and other insects that have the potential to sting or bite an attacker, a phenomenon known as Batesian mimicry. Not only do the mimics have the same colours and furry textures as their models, but some also move and behave in a similar way and may even produce a bee-like hum. This may cause confusion not only to potential predators, but also budding entomologists attempting to identify and separate bumblebees from other similar-looking but completely unrelated insects.

Most bumblebee mimics belong to the order Diptera, the flies, comprising of around 7,000 species in Britain and Ireland. Within the Diptera, the majority of bumblebee mimics belong to the family Syrphidae, the hoverflies, and some of these have several colour forms that mimic different species of bumblebee. However, on closer examination – and as the name Diptera (meaning 'two wings') suggests – flies have just one pair of wings, compared to a bee's two pairs, and their antennae are short and simple rather than long and sharply angled. Flies also have much larger eyes, with those of males often meeting in the middle.

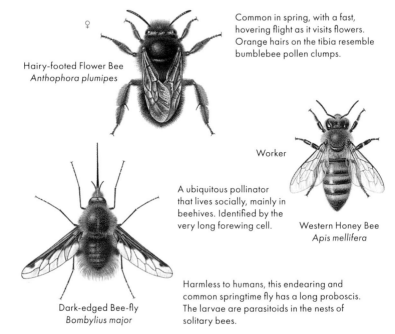

♀

Common in spring, with a fast, hovering flight as it visits flowers. Orange hairs on the tibia resemble bumblebee pollen clumps.

Hairy-footed Flower Bee
Anthophora plumipes

Worker

A ubiquitous pollinator that lives socially, mainly in beehives. Identified by the very long forewing cell.

Western Honey Bee
Apis mellifera

Dark-edged Bee-fly
Bombylius major

Harmless to humans, this endearing and common springtime fly has a long proboscis. The larvae are parasitoids in the nests of solitary bees.

A large, furry bumblebee mimic. Rare, found in ancient pine forests in the Scottish Highlands.

Bumblebee Robberfly
Laphria flava

Bear Syrph
Eriozona syrphoides

A rare hoverfly, found in coniferous and mixed woodland mainly in Wales and north-west Britain.

A fairly common hoverfly, usually found close to umbellifers, where its larvae feed in the roots.

Bumblebee Blacklet
Cheilosia illustrata

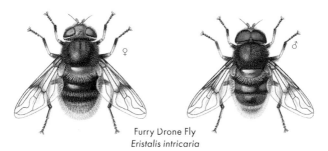

Furry Drone Fly
Eristalis intricaria

A drone fly that is variable in colour, with specimens found in the spring usually being darker. Occurs in wetlands and damp woodlands throughout Britain and Ireland.

Various colour forms mimic different species of bumblebee. Widespread and common in gardens, where the larvae feed in the bulbs of daffodils and narcissus.

Large Narcissus Fly
Merodon equestris

Different colour forms mimic red-tailed bumblebees and white-tailed bumblebees. Common and widespread in many habitats, including gardens.

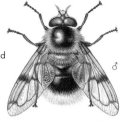

Bumblebee Hoverfly
Volucella bombylans

A mimic of carder bees. Found locally in peaty woodlands, mainly in the north and west of Britain and Ireland.

Furry Peat Hoverfly
Sericomyia superbiens

A bearfly that occurs in two colour forms, these resembling the Tree Bumblebee (see pages 40–41) and Common Carder Bee (see pages 62–63). Has obscure dark bands on the thorax. Widespread but not common; found in ancient woodland, where its larvae live in decaying wood.

Dimorphic Bearfly
Criorhina berberina

A bearfly that resembles carder bumblebees, with prominent pale tufts at the sides of the thorax and abdomen. Widespread but not common; found in old woodland and hedgerows, and often seen on flowers of Hawthorn and Bramble.

♂

Buff-tailed Bearfly
Criorhina floccosa

Occurs in two colour forms, with either a red or white tail. The hind tibia is swollen and the tarsi curved. Local in ancient woodlands, mainly in southern Britain. Feeds from flowers of various *Prunus* species, willows and Hawthorn.

♂

Early Bearfly
Criorhina ranunculi

Resembles the White-tailed Bumblebee aggregate. Found mainly in southern England. Very rare in ancient woodland such as the New Forest.

Bumblefly *Pocota personata*

A large, spiny black and orange fly. Occurs in woods and heathland. Locally widespread throughout Britain and Ireland. Parasitises moth larvae.

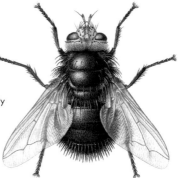

Giant Tachinid Fly *Tachina grossa*

Females lay their eggs in the nostrils of Red Deer (*Cervus elaphus*). Widespread in northern Scotland.

Deer Bot-fly
Cephenemyia auribarbis

Flies by day and feeds on the petals of Bramble and Wild Thyme. Larvae feed on decaying wood. Uncommon; found mostly in Wales and the Scottish Highlands.

Bee Chafer
Trichius fasciatus

A local day-flying moth with a fast, darting flight. Found on downland and moors throughout Britain and Ireland. Its caterpillars feed on scabiouses.

Narrow-bordered Bee Hawk-moth
Hemaris tityus

A fast day-flying hawk-moth, found mainly in woodland rides in the southern half of England and Wales. Its caterpillars feed on Honeysuckle.

Broad-bordered Bee Hawk-moth
Hemaris fuciformis

Glossary

Abdomen The posterior of the three main body sections.

Antenna (pl. antennae) The sensory appendage at the front of the head.

Basitarsus (pl. basitarsi) The first and largest of the five tarsal segments.

Caste A form of social insect, either a queen, worker or male.

Clypeus The lower part of the face below the eyes.

Cocoon A silken chamber in which the larva pupates.

Compound eyes The two large eyes made up from many tiny individual lenses.

Corbicula (pl. corbiculae) The concave, shiny, hair-fringed part of the hind tibia where pollen is collected by female bumblebees; also called the pollen basket.

Diptera The order of insects comprising the flies.

Exoskeleton The hard outer cuticle of an insect or other arthropod's body.

Femur The third leg segment, equivalent to the human thigh.

Flagellum (pl. flagella) The final 10 (female) or 11 (male) segments of the antennae.

Hymenoptera The order of insects comprising the sawflies, wasps, bees and ants.

Keel A raised ridge on sternite 6 of some female bumblebees.

Mandible The jaws at the lower part of the head.

Kleptoparasite A parasite that takes over the nest of another species, e.g. a cuckoo bumblebee.

Machair Low-lying flower-rich grassland on islands and coastal parts of north-west Scotland and Ireland.

Melanic A black or very dark form of a bumblebee.

Ocellus (pl. ocelli) Three small, simple, light-sensitive eyes on the top of the head.

Parasitoid A parasite that lives on or in its host, eventually killing it.

Pedicel The small second antennal segment between the scape and the flagellum.

Pocket maker A bumblebee that stores pollen in wax pots alongside its larvae.

Pollen basket See 'Corbicula'.

Pollen storer A bumblebee that stores pollen in wax pots away from its larvae.

Proboscis The bumblebee tongue.

Scape The first and longest antennal segment.

Scutellum The rear dorsal part of the thorax.

Sternite The underside segments of the abdomen.

Tarsus (pl. tarsi) The last five segments of each leg.

Tegula (pl. tegulae) The cup-shaped plate at the base of each wing.

Tergite The upperside segments of the abdomen.

Thorax The middle body section where the wings and legs are attached.

Tibia (tibiae) The fourth leg segment between the femur and tarsi.

Usurpation The process whereby a true bumblebee queen takes over the established nest of another queen.

Gardens and bumblebees

As the decline in bumblebees and many other forms of wildlife has continued in the wider countryside over the last 70 years, so the awareness of the importance of gardens as wildlife habitats has grown. This is reflected in the plethora of wildlife gardening books and magazines, and the popularity of recording garden wildlife through schemes such as the Royal Society for the Protection of Birds' Big Garden Birdwatch and Butterfly Conservation's Garden Moth Scheme. The British Isles has more than 20 million gardens covering an area equal in size to one-fifth of Wales – several times larger than all our nature reserves combined and extending to more than 4,000sq km. It could therefore be argued that, with no two gardens exactly alike, these habitats are as important to wildlife as are our nature reserves, particularly to the more widespread invertebrates. It is also the case that gardens often have a greater diversity of wildlife and microhabitats than exists in the surrounding open farmed countryside. This huge diversity of life contained in gardens was reflected in Jennifer Owen's *Wildlife of a Garden: A Thirty-year Study* (2010), which looked at an average suburban garden in central England and revealed an astonishing 1,100-plus invertebrates, including 13 species of bumblebee.

Encouraging and increasing the diversity of wildlife in our gardens doesn't necessarily require drastic action, and small changes can often be hugely beneficial. Even small outside areas such as balconies can benefit from having a small log pile or pots containing plants for pollinators, and the increase in the number of species attracted to such features will give great satisfaction to the gardener. The main thing to consider when encouraging bumblebees and other invertebrates into gardens is variety, not just of plants but also of microhabitats. Remember that all animals, even closely related species, often have particular requirements for feeding, resting, overwintering and breeding. In addition to a good range of plants supplying nectar and pollen for sustenance, bumblebees need less tidy, sheltered areas in which to breed and seek refuge. Lawns or sections of lawns will benefit from less frequent mowing, allowing plants such as clovers, Selfheal and Dandelion to flower, and of course we should always avoid using herbicides and insecticides. Over-tidying gardens in winter should also be avoided, as it may lead to the loss of suitable habitats for hibernating invertebrates.

As bumblebee tongue length varies from species to species, a variety of plants should be included in the garden. Plant a range, from those with flat, open heads such as sedums and sunflowers, to those with flowers that have a long corolla, such as Foxglove and Comfrey. It is best to avoid plants that have double or multi-petalled flowers, such as some roses and chrysanthemums, as these prevent access to pollen and nectar, or those that produce very little pollen or nectar, such as petunias and pelargoniums. Further guidance on what to plant is available through schemes such as the Royal Horticultural Society's 'Plants for Pollinators' list, and the Bumblebee Conservation Trust's 'Bee kind' tool.

Most gardens are made up of a 30:70 proportion of native to non-native plants. Bumblebees and all other pollinators do not discriminate between these and will visit either, providing there is an ample and accessible source of nectar and pollen. However, when planting for pollinators, a bias towards native plants is generally the best choice. As bumblebees need food from spring until autumn and sometimes in the winter months, include garden plants that provide a continuous succession of flowers from February until October. For species like the Buff-tailed Bumblebee, which sometimes forages in the winter, include plants such as *Mahonia*, Winter Jasmine and hellebores to help their survival.

Below is a selected list of plants that are regarded as bumblebee favourites, although they will also find many other species attractive.

Native and non-native trees and shrubs

Apple *Malus domestica*
Bird Cherry *Prunus cerasifera*
Blackthorn *Prunus spinosa*
Common Gorse *Ulex europaeus*
Flowering Currant *Ribes sanguineum*
Garden Lavender *Lavandula angustifolia*
Goat Willow *Salix caprea*
Hawthorn *Crataegus monogyna*
Japanese Mahonia *Mahonia japonica*
Wayfaring-tree *Viburnum lantana*
Winter-flowering Honeysuckle *Lonicera fragrantissima*
Winter Jasmine *Jasminum nudiflorum*

Native flowering plants

Bell Heather *Erica cinerea*
Betony *Stachys officinalis*
Bilberry *Vaccinium myrtillus*
Bramble *Rubus fruticosus* agg.
Bugle *Ajuga reptans*
Calamint *Calamintha nepeta*
Comfrey *Symphytum officinale*
Common Bird's-foot Trefoil *Lotus corniculatus*
Common Knapweed *Centaurea nigra*
Creeping Thistle *Cirsium arvense*
Dandelion *Taraxacum officinale*
Devil's-bit Scabious *Succisa pratensis*
Field Scabious *Knautia arvensis*
Foxglove *Digitalis purpurea*
Ground-ivy *Glechoma hederacea*
Honeysuckle *Lonicera periclymenum*

Ivy *Hedera helix*
Kidney Vetch *Anthyllis vulneraria*
Lesser Calamint *Clinopodium calamintha*
Lungwort *Pulmonaria officinalis*
Marsh Woundwort *Stachys palustris*
Primrose *Primula vulgaris*
Raspberry *Rubus idaeus*
Red Bartsia *Odontites vernus*
Red Clover *Trifolium pratense*
Selfheal *Prunella vulgaris*
Spiked Speedwell *Veronica spicata*
Teasel *Dipsacus fullonum*
Tufted Vetch *Vicia cracca*
Viper's Bugloss *Echium vulgare*
White Clover *Trifolium repens*
White Dead-nettle *Lamium album*
Wild Marjoram *Origanum vulgare*
Wild Thyme *Thymus drucei*
Yellow Rattle *Rhinanthus minor*

Non-native flowering plants

Anise Hyssop *Agastache foeniculum*
Black-eyed Susan *Rudbeckia fulgida*
Blue Eryngo *Eryngium planum*
Borage *Borago officinalis*
Chives *Allium schoenoprasum*
Common Snowberry *Symphoricarpos albus*
Dwarf Catmint *Nepeta racemosa*
Hyssop *Hyssopus officinalis*
Lamb's Ear *Stachys byzantina*
Macedonian Scabious *Knautia macedonica*
Michaelmas Daisy *Aster amellus*
Miss Willmott's Ghost *Eryngium giganteum*
New York Aster *Symphyotrichum novi-belgii*
Purple Top *Verbena bonariensis*
Rosemary *Salvia rosmarinus*
Sneezeweed *Helenium autumnale*

Further reading and resources

Books

Benton, T 2006 *Bumblebees*. Collins New Naturalist Library

Comont, R 2017 *RSPB Spotlight Bumblebees*. Bloomsbury Publishing

Edwards, M and Jenner, M 2018 *Field Guide to the Bumblebees of Great Britain and Ireland*. 3rd edition. Ocelli Limited

Falk, S and Lewington, R 2015 *Field Guide to the Bees of Great Britain and Ireland*. Bloomsbury Publishing

Gammans, N, Comont, R, Morgan, S C and Perkins, G 2018 *Bumblebees: An Introduction*. Bumblebee Conservation Trust

Goulson, D 2009 *Bumblebees: Behaviour, Ecology and Conservation*. 2nd edition. Oxford University Press

Goulson, D 2021 *Gardening for Bumblebees*. Square Peg

Owens, N 2020 *The Bumblebee Book*. Pisces Publications

Prŷs-Jones, O E and Corbet, S A 2011 *Bumblebees*. 3rd edition. Naturalists' Handbook 6. Pelagic Publishing

Sladen, F W L 1912 *The Humble-bee: Its Life-history and How to Domesticate It*. Macmillan

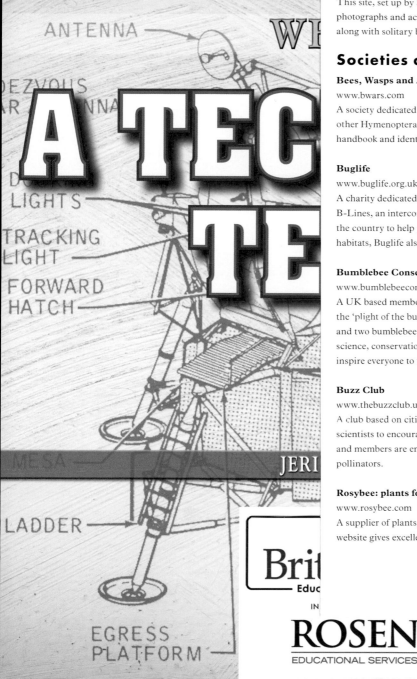

Online resources

British Bees on Flickr
http://tinyurl.com/nrywslu
This site, set up by Steven Falk, contains the most extensive collection of excellent photographs and accompanying texts on all bumblebees on the British and Irish lists, along with solitary bees and many other insect orders.

Societies and organisations

Bees, Wasps and Ants Recording Society (BWARS)
www.bwars.com
A society dedicated to gathering distributional and biological data on bumblebees and other Hymenoptera. Membership includes a twice-yearly newsletter, a members' handbook and identification workshops throughout the year.

Buglife
www.buglife.org.uk
A charity dedicated to promoting the importance of invertebrates, including the project B-Lines, an interconnected series of flower-rich insect pathways running throughout the country to help pollinating insects. As well as campaigning to preserve threatened habitats, Buglife also offers advice on how to attract and identify garden invertebrates.

Bumblebee Conservation Trust (BBCT)
www.bumblebeeconservation.org
A UK based membership charity that was established because of serious concerns about the 'plight of the bumblebee'. In the last 80 years bumblebee populations have crashed and two bumblebee species have become extinct in the UK. The Trust carries out science, conservation, and engagement activities, influences policymakers, and hopes to inspire everyone to take action to help bumblebees.

Buzz Club
www.thebuzzclub.uk
A club based on citizen science, which brings together naturalists, gardeners and scientists to encourage further understanding of garden wildlife. Membership is free and members are encouraged to take part in projects that help bumblebees and other pollinators.

Rosybee: plants for pollinators
www.rosybee.com
A supplier of plants, attractive to bees, grown without the use of peat or pesticides. The website gives excellent, well-researched information on which plants attract which bees.

Published in 2015 by Britannica Educational Publishing (a trademark of Encyclopædia Britannica, Inc.) in association with The Rosen Publishing Group, Inc.
29 East 21st Street, New York, NY 10010

Distributed exclusively by Rosen Publishing.
To see additional Britannica Educational Publishing titles, go to rosenpublishing.com.

First Edition

Britannica Educational Publishing
J. E. Luebering: Director, Core Reference Group
Mary Rose McCudden: Editor, Britannica Student Encyclopedia

Rosen Publishing
Hope Lourie Killcoyne: Executive Editor
Nelson Sá: Art Director
Michael Moy: Designer
Cindy Reiman: Photography Manager
Karen Huang: Photo Researcher

Library of Congress Cataloging-in-Publication Data

Freedman, Jeri.
What is a technical text?/Jeri Freedman. — First edition.
    pages cm. — (The Britannica common core library)
Includes bibliographical references and index.
ISBN 978-1-62275-672-8 (library bound) — ISBN 978-1-62275-673-5 (pbk.) — ISBN 978-1-62275-674-2 (6-pack)
1. Technical literature — Juvenile literature. 2. Technical writing — Juvenile literature. I. Title.
T10.7F74 2015
600 — dc23
                                                                                            2014020385

Manufactured in the United States of America

Photo credits:

Cover (background) © iStockphoto.com/marcoventuriniautieri; cover (hands and tablet) © iStockphoto.com/Anatoliy Babiy; cover (tablet screen) © iStockphoto.com/elgol; p.1, interior pages background image Diagram of the Apollo Lunar Module. Courtesy of NASA History Office; p. 4 Mark Hall/The Image Bank/Getty Images; pp. 5, 17 Encyclopædia Britannica, Inc.; pp. 6, 9, 12, 13, 22, 24, 27 © Rosen Publishing; p. 7 © iStockphoto.com/Timothy Masters; p. 8 Mikko Lemola/Shutterstock.com; p. 14 Andresr/Shutterstock.com; p. 15 A and N photography/Shutterstock.com; p. 16 Christopher Futcher/E+/Getty Images; p. 18 Mathew Ward/Dorling Kindersley/Getty Images; p. 19 Diana Haronis/Moment/Getty Images; p. 20 oliveromg/Shutterstock.com; p. 21 Thierry Foulon/PhotoAlto Agency RF/Getty Images; p. 25 AP Images for Macy's; p. 29 Hurst Photo/Shutterstock.com.

CONTENTS

What Is a Technical Text?

A technical text is a special type of nonfiction. It is meant to teach readers something new about a topic. It often instructs readers on how to perform a task.

Technical texts can be found everywhere. Instruction booklets or leaflets that come with household items are technical texts. These can include instructions for how to play a game.

Labels on machinery that contain safety warnings and operating instructions are technical texts. Recipes and other useful how-to information are common technical texts found on the Internet or in books.

A technical text often tells how to do something, such as how to set up a tent.

Labels are words or names on something to describe or show what it is.

A technical text may explain how to perform an experiment or make a stuffed animal. It can describe how to fit together the parts of a robot or build a tree house. Technical texts are made for readers who are learning something new. Therefore, they must present information so that it is easy to understand.

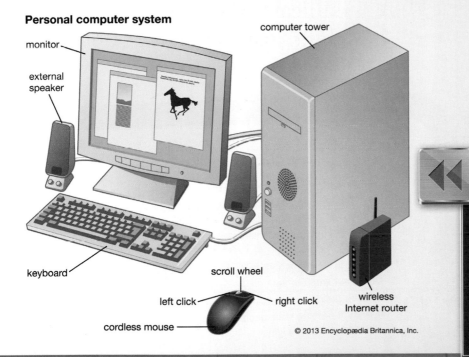

Personal computer system

computer tower

monitor

external speaker

keyboard

scroll wheel

left click

right click

cordless mouse

wireless Internet router

© 2013 Encyclopædia Britannica, Inc.

Labels show the parts of a personal computer system. A technical text can explain how to put together the parts of a personal computer system.

What Qualities Does a Technical Text Have?

To help people learn new skills and ideas, technical writing must have the following qualities.

Clear and Compact Writing

Technical texts use language that is easy to understand. Short, simple words and sentences are used. Only necessary information is included. The words are organized to help readers

A technical text uses bullets, or dots, in a list to make clear the most important points.

Safe Strength Training

As with any activity, strength training can result in injury. You can reduce your risk of injury by taking part in a well-supervised strength-training program with a qualified adult. Listening and following instructions will also help prevent injury, as will the following tips:

- Be sure to wear appropriate clothing and closed-toe athletic shoes.
- Start your workout with a five- to ten-minute warm-up. Walking, running in place, or jumping rope are good options.
- If you are training with machines, make sure they have been adjusted for your height.

understand the topic. Information is given in the order that it is needed.

> **Graphs** are drawings that show mathematical information with lines, shapes, and colors. **Diagrams** are drawings that explain or show the parts of something.

Helpful Presentation

Technical texts often use design elements to highlight important points. These elements can include bullet points, numbered lists, headings, boldface, underline, and blank space. Visual aids, such as tables, charts, maps, **graphs**, and other **diagrams,** are also used to illustrate important ideas from the text.

Diagrams help to guide a person in putting together an object, such as a birdhouse.

Reader Awareness

The goal of every technical text is to teach the reader something specific. To do this, technical writers must understand what their audience needs to know. Information that is new to the reader is defined.

Technical texts are not meant to entertain readers. The writing must have a serious tone, or feeling. The reader will know to pay close attention to the information in the text.

Correct Information

People use technical texts to do important jobs. Some of these jobs, like fixing a car, can produce dangerous conditions when they are not done correctly. Errors in technical

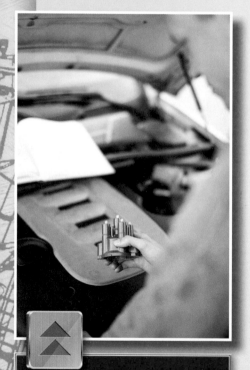

Car repair manuals are instruction booklets that mechanics often use when they fix cars.

WHAT QUALITIES DOES A TECHNICAL TEXT HAVE?

Technical Writers

The World Record Paper Airplane Book (2006), by engineers Ken Blackburn and Jeff Lammers, teaches readers how to fold and fly different types of paper airplanes—as well as why airplanes fly and crash.

writing can cause physical injury, damage to property, or misunderstandings.

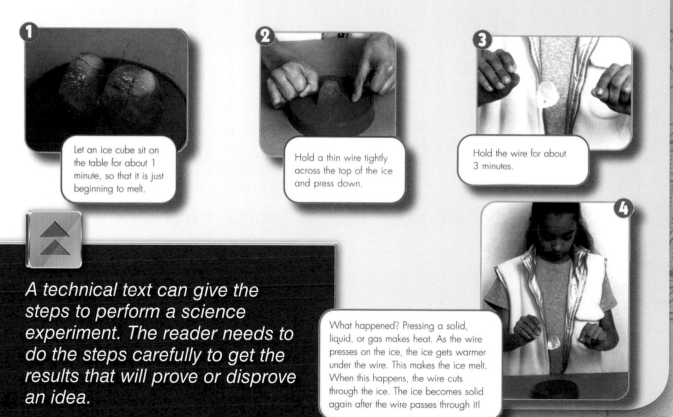

1 Let an ice cube sit on the table for about 1 minute, so that it is just beginning to melt.

2 Hold a thin wire tightly across the top of the ice and press down.

3 Hold the wire for about 3 minutes.

4 What happened? Pressing a solid, liquid, or gas makes heat. As the wire presses on the ice, the ice gets warmer under the wire. This makes the ice melt. When this happens, the wire cuts through the ice. The ice becomes solid again after the wire passes through it!

A technical text can give the steps to perform a science experiment. The reader needs to do the steps carefully to get the results that will prove or disprove an idea.

Technical Text in Use

The following examples show how technical texts present information differently for different purposes.

A Scientific Study

A scientific study examines events to explain their effects on the world. It often includes mathematical information. Graphs, or charts, are used to compare amounts of things. The most common kinds of graphs are circle graphs (or pie charts), **bar graphs**, and line graphs. The passage below uses a bar graph and a

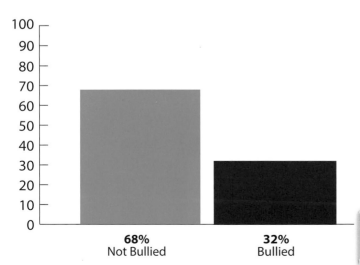

Percentage of Students Bullied and Not Bullied

68% Not Bullied

32% Bullied

A bar chart shows the percent of students who were bullied and not bullied while at school.

pie chart to show how bullying affects groups of students.

According to the National Center for Education Statistics, in 2007, 32 percent of students from 12 to 18 years old reported having been bullied at school during the school year. Of those, 11 percent said that they were pushed, shoved, tripped, or spit on. Four percent said that someone tried to make them do things they did not want to do or that their property was destroyed on purpose.

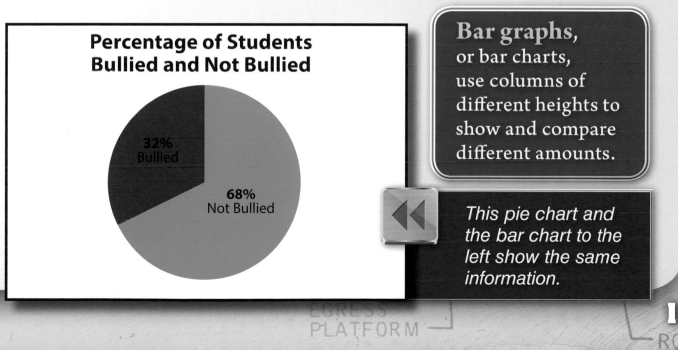

Percentage of Students Bullied and Not Bullied

32% Bullied

68% Not Bullied

Bar graphs, or bar charts, use columns of different heights to show and compare different amounts.

This pie chart and the bar chart to the left show the same information.

A Scientific Experiment

A scientific experiment involves a series of actions and observations to test an idea. Below is an experiment dealing with the evaporation of water from a carrot. Evaporation is a process through which a liquid changes to a gas. Vegetables, fruit, and other plant products often contain liquid water.

1. *Use a scale to weigh a fresh carrot that has been cut into several small pieces. Record the total weight of all the carrot pieces.*

Following specific steps and making careful observations while doing a scientific experiment allow you to test an idea.

2. Set the carrot pieces out on a paper plate for several days.

3. Weigh the carrot pieces a second time and record their total weight.

4. Compare the new weight of the carrot to the weight you recorded earlier.

Steps in an experiment are usually numbered to help readers do them in the correct order.

Technical Writers

Writer Sy Montgomery and illustrator Nic Bishop show readers a scientific study in the book *Kakapo Rescue: Saving the World's Strangest Parrot* (2010).

Let's Compare

Studies and experiments are two methods used by scientists. Both provide the reader with **data**, or pieces of information.

The bullying study reports findings to the reader. It does not expect the reader to do part of the study or any task related to the study. All of the information that the reader is expected to consider is included in the text. The study also includes visual aids to help the reader understand

People often gather information when making a scientific study. That information can be examined and used to explain ideas about the world.

Data are facts or measurements used to explain ideas.

By following the steps for doing an experiment, scientists can prove whether or not their ideas about what should happen are true.

the mathematical information.

The second example provides the steps for the reader to conduct an experiment. The steps involved in the experiment are numbered so that the reader can do them in the correct order. Readers are expected to gather and interpret the information from the experiment themselves.

How to Play Volleyball

The following passage describes the main rules of volleyball.

Play begins when one player serves. A player serves by hitting the ball over to the receiving team's side of the net. The receiving team tries to return the serve, or hit the ball back over the net. Players may hit the ball with any part of the body above the waist as long as the ball is clearly hit and not held. Each team may hit the ball up to three times before sending it back over the net. The serving team

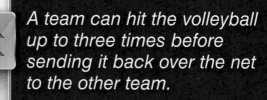

A team can hit the volleyball up to three times before sending it back over the net to the other team.

Technical Writers

Writers Abigael McIntyre, Sandra Giddens, and Owen Giddens explain to readers how to play and train for the game of volleyball in *An Insider's Guide to Volleyball* (2015).

scores a point if the other team fails to return the ball or hits it out of bounds. When the serving team fails to return the ball or hits it out of bounds, it loses the serve to the other team. A winning score is at least fifteen points, with at least a two-point lead.

VOLLEYBALL COURT

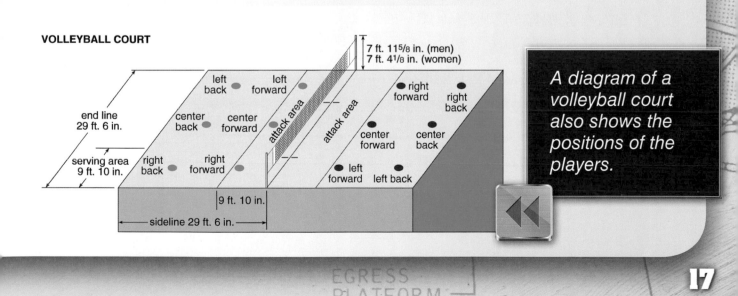

A diagram of a volleyball court also shows the positions of the players.

How to Play Checkers

Games often include **manuals** or instruction leaflets that explain all of their rules. The following passage describes some of the rules of playing checkers.

The game is played by two people. The players move disc-shaped pieces around a game board that has alternating squares of black and some other color.

Step 1: The players place twelve game pieces each on the black squares in three rows at opposite sides of the board.
Step 2: One player starts by moving one of the pieces

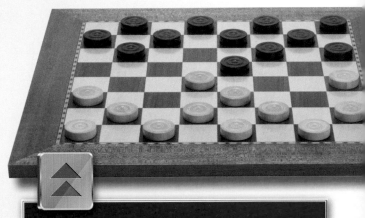

A checkerboard is marked with squares of two different colors. Each player uses different colored game pieces called checkers.

Manuals are small books that give useful information about how to use something.

in the row farthest from the edge of the board. The player can move the piece only one square at a time. The player can only move forward, and the move must be diagonal so that the piece ends up on another black square.

The game of checkers is played by two people. Players can move only one game piece at a time.

Let's Compare

The passages about volleyball and checkers give important information about how to play each game and use illustrations to highlight the text. However, each set of directions is organized differently.

The passage about how to play volleyball gives only the basic rules of the game. It is written for an audience that is unfamiliar with the sport. After reading the passage, the reader will be able to start playing. Although all of the information is important for the reader to know, the reader is not required to follow the instructions in a particular order.

The text example for how to play volleyball gives the basic rules for playing the game. Players do not have to follow the directions in a particular order after serving the ball.

The passage about how to play checkers presents information in an order that must be followed. The first two sentences appear first because they describe what is needed for any game to take place: two players, a checkerboard, and game pieces. Step 1 must be done before Step 2.

In the game of checkers, players have to complete Step 1 before going on to Step 2 when playing the game.

A Recipe for Hot Chocolate

This recipe for Mexican hot chocolate includes a **safety warning** to young people about using a stove and getting a grown-up to help.

Ingredients
½ cup (120 ml) sugar
¼ cup (60 ml) cocoa
¼ teaspoon salt
1 tablespoon all-purpose flour
1 teaspoon cinnamon
1 cup (240 ml) cold water
4 cups (1 l) milk
2 teaspoons vanilla extract

1. Place the sugar, cocoa, salt, flour, cinnamon, and water in a large pot. Place the pot over low heat, and stir until the mixture dissolves into the water.

Recipes are technical texts that help people create their own tasty treats.

Instructions often contain a **safety warning**. These warnings are often placed in a box, colored circle, or other shape so that they are easily noticed. It is important to follow safety warnings so that no one gets hurt.

2. *After the mixture is dissolved, turn the stove up to medium high. Stir the mixture until it boils. Boil for three to five minutes, stirring all the time.*
3. *Add the milk. Continue to stir until the milk is hot but not boiling.*
4. *Just before the milk boils, remove the pot from the heat. Stir in the vanilla extract, and serve!*

!

When cooking, you should always have an adult with you in the kitchen to help. Many of the tools used to prepare these recipes and others can be dangerous. Always be very careful when using a knife or a stove.

Take special care to notice and read any safety warnings that appear in technical texts. Ask an adult that you know to read the warning label and help you complete the task.

How to Make a Mask

This example shows the first few steps in how to make a simple cat mask.

1. *Draw a circle around a plate to make a circle of cardboard about 8 inches (20 cm) across. Cut out the circle.*
2. *Cut out eyeholes the same distance apart as your own eyes. Cut a flap for your nose.*
3. *With a glue stick, stick blue paper to the cardboard to cover it. Stick pink eyes cut from paper around the eyeholes. Cut out cardboard ears and stick them on.*

This photograph helps readers see how each numbered step adds to the finished cat mask. It also shows readers what the finished mask should look like.

Technical Writers

Martha Stewart is famous for her homemaking arts and crafts. Her business produced a number of arts and crafts books, including *Martha Stewart's Favorite Crafts for Kids: 175 Projects for Kids of All Ages to Create, Build, Design, Explore, and Share.*

4. *Cut out a mouth, tongue, and some little circles from paper and stick them on. Glue on pieces of drinking straw for whiskers.*

Martha Stewart explains how to make holiday popcorn balls. Stewart enjoys cooking and crafts and has written many books and articles on those subjects.

Let's Compare

Recipes and craft directions usually follow a similar style of **presentation**. Materials or ingredients are listed first. Then the directions for how to make the dish or craft follow in numbered steps. Presenting the directions in this way is important because steps need to be done in the correct order.

Recipes and craft directions often use visual aids differently. Craft directions usually include illustrations or photographs that show how to do each step described in the text. Recipes normally do not illustrate each numbered step. They sometimes show a photograph of the raw ingredients, especially if they are unusual or hard to find in stores. Most often recipes will have a photograph of the finished dish only. Both recipes and craft instructions may contain warnings about any possible dangers involved with the task.

The **presentation** of recipes or directions means the way in which the information is shown, described, or explained.

Ingredients

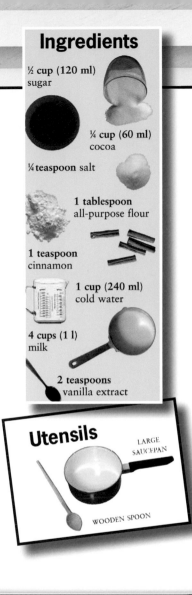

½ cup (120 ml)
sugar

¼ cup (60 ml)
cocoa

¼ teaspoon salt

1 tablespoon
all-purpose flour

1 teaspoon
cinnamon

1 cup (240 ml)
cold water

4 cups (1 l)
milk

2 teaspoons
vanilla extract

Utensils

LARGE
SAUCEPAN

WOODEN SPOON

1 Place the sugar, cocoa, salt, flour, cinnamon, and water in a large saucepan. Place the saucepan over low heat and stir until the mixture **dissolves** into the water.

2 After the mixture is dissolved, turn the stove up to medium high. Stir the mixture until it boils. Boil for 3 to 5 minutes, stirring all the time.

3 Add the milk. Continue to stir until the milk is hot but not boiling.

4 Just before the milk boils, remove the saucepan from the heat. Stir in the vanilla extract, and serve!

Some recipes include visual aids for every ingredient, tool, and step that is needed. Here are visual aids for the Mexican hot chocolate recipe.

27

Write Your Own Technical Text

Now it is your turn to write a technical text. Your text should tell someone how to make an object, complete an experiment, or do a task. Follow these steps:

1. Pick your topic. Choose something that you already know how to do well and that someone else might enjoy learning how to do.
2. Choose your audience. Will your audience know much about the topic? What will they need you to define?
3. Write an introduction to the activity that includes important points and definitions.
4. Write the steps to do the activity.
5. Draw or print pictures to help the reader understand the directions.

6. Put what you wrote aside for a day. Next reread it as if you were the audience. Is any information missing, unnecessary, or incorrect?

7. Have some fun! Ask someone you know to follow your directions. Can he or she do it?

By watching classmates follow your directions, you will learn whether your technical text is clear and correct.

charts Information presented in the form of tables, diagrams, or illustrations.

diagrams Drawings that explain or show the parts of something.

engineers People who design and build things such as bridges and roads.

graphs Drawings that show mathematical information with lines, shapes, and colors.

nonfiction Writing that is about facts or real events.

observations Acts of careful watching and listening to gather information.

tables Information displayed in forms that often include rows and columns.

technical Relating to the practical use of machines or science in industry, medicine, and other fields.

visual aids Illustrations or diagrams that make a text easier to understand.

Books

Blackburn, Ken, and Jeff Lammers. *The World Record Paper Airplane Book.* New York, NY: Workman Publishing, 2007.

Martha Stewart Living editors. *Martha Stewart's Favorite Crafts for Kids: 175 Projects for Kids of All Ages to Create, Build, Design, Explore, and Share.* New York, NY: Potter Craft, 2013.

McIntyre, Abigael, Sandra Giddens, and Owen Giddens. *An Insider's Guide to Volleyball.* New York, NY: Rosen Publishing Group, 2015.

Montgomery, Sy. *Kakapo Rescue: Saving the World's Strangest Parrot.* New York, NY: Houghton Mifflin Books for Children, 2010.

Oxlade, Chris. *The Science and History Project Book.* Helotes, TX: Armadillo Books, 2013.

Ward, Karen. *Fun with Mexican Cooking.* New York, NY: Rosen Publishing Group, 2010.

Williams, Zella. *Experiments with Solids, Liquids, and Gases.* New York, NY: Rosen Publishing Group, 2007.

Websites

Because of the changing nature of Internet links, Rosen Publishing has developed an online list of websites related to the subject of this book. This site is updated regularly. Please use this link to access the list:

http://www.rosenlinks.com/BCCL/Tech